AGRICULTURE ISSUES AND POLICIES

COOPERATIVES

A FEASIBILITY STUDY GUIDE AND EDUCATION EFFORTS OF FARMERS

AGRICULTURE ISSUES AND POLICIES

Additional books in this series can be found on Nova's website under the Series tab.

Additional E-books in this series can be found on Nova's website under the E-book tab.

BUSINESS ISSUES, COMPETITION AND ENTREPRENEURSHIP

Additional books in this series can be found on Nova's website under the Series tab.

Additional E-books in this series can be found on Nova's website under the E-book tab.

AGRICULTURE ISSUES AND POLICIES

COOPERATIVES

A FEASIBILITY STUDY GUIDE AND EDUCATION EFFORTS OF FARMERS

JACOB ROGERS
AND
VINCENT K. GRANT
EDITORS

Nova Science Publishers, Inc.
New York

For permission to use material from this book please contact us:
Telephone 631-231-7269; Fax 631-231-8175
Web Site: http://www.novapublishers.com

Additional color graphics may be available in the e-book version of this book.

Library of Congress Cataloging-in-Publication Data

ISBN: 978-1-62081-251-8

Published by Nova Science Publishers, Inc. † New York

CONTENTS

PREFACE

This book provides a national perspective of cooperative education initiatives by farmer cooperatives. The analysis examines whom cooperatives educate, what resources they use, what topics they focus on, where they source their programs and materials, whether they feel there are enough resources available for education, and what educational topics they feel directors most need to fulfill their duties. A cooperative feasibility study guide is also outlined with a focus on the cooperative business development process.

Chapter 1- This guide provides rural residents with information about cooperative development feasibility studies. It defines the feasibility study and discusses their necessity and limitations. First steps in feasibility study development are described and key actions, including important components of a comprehensive study, are detailed. Also offered are criteria for selecting and working with consultants, information for developing assumptions, and study assessment factors.

Chapter 2- This report describes farmer cooperatives' education initiatives by providing information on whom they are educating and how, topics covered, their sources for educational materials and programs, and what educational topics they feel directors most need. The degree of cooperative education provided is reported for respondents overall and by cooperative type, size, and location.

In: Cooperatives
Editors: J. Rogers and V. K. Grant
ISBN: 978-1-62081-251-8

Chapter 1

VITAL STEPS: A COOPERATIVE FEASIBILITY STUDY GUIDE*

John W. Brockhouse, Jr. and James J. Wadsworth

ABSTRACT

This guide provides rural residents with information about cooperative development feasibility studies. It defines the feasibility study and discusses their necessity and limitations. First steps in feasibility study development are described and key actions, including important components of a comprehensive study, are detailed. Also offered are criteria for selecting and working with consultants, information for developing assumptions, and study assessment factors.

ACKNOWLEDGMENTS

John W. Brockhouse, Jr., and James J. Wadsworth completed the revision of this guide. The report was first authored by James Matson in 2000 while he was employed with Rural Development. The authors acknowledge revision contributions from Bruce Pleasant, with USDA Rural Development in North

* This is an edited, reformatted and augmented version of United States Department of Agriculture, Rural Business-Cooperative Service, Service Report 58, dated December 2010.

Carolina, and James Matson, now a private consultant based in South Carolina.

Publications and information are also available on the Internet. The Cooperative Programs Web site is at: http://www.rurdev.usda.gov/rbs/coops /csdir.htm

THE COOPERATIVE BUSINESS DEVELOPMENT PROCESS

In most cases, developing a business is a complex undertaking that consists of a number of stages. In cooperative development, following a step-by-step deliberative process is important because of the need to review progress and obtain input from potential members after each step before making a decision about whether to proceed with the project. The report "How to Start a Cooperative" (RBS CIR 7), presents a 16-step sequence of events recommended for creating a cooperative business (see Appendix A).

This guide elaborates on the seventh of the 16- step event sequence: conducting a feasibility analysis. A feasibility study is an integral part of cooperative business development. Conducting a feasibility study is not a theoretical research project, but rather is a focused study for a specific group that wants to explore forming a cooperative business.

Figure 1 condenses the sequence of events of the cooperative development process into four stages and shows how the feasibility study fits within that process. The feasibility study occurs during the deliberation stage when the focus is on whether to proceed with a project. (It should be noted that the steps in Figure 1 are sometimes approached and completed in a different order than shown. The specific order of the steps taken in a development project will be contingent on the type of venture being explored and the wishes of those involved.)

The amount of time required in the cooperative business development process depends on the complexity of the proposed business and the attributes of the group involved. Typically, it takes 1 to 2 years for a cooperative development project to be completed, although there are some cases where projects are completed faster and others that take longer.

The time needed to complete the feasibility study step varies widely from project to project, again depending on the characteristics of the group requesting the study, as well as the specific aspects of the venture, such as technological complexity, project scale, marketing conditions, member involvement, and financial planning factors, etc. However, a good rule of thumb for the feasibility analysis step for most development projects is 3 to 6 months.

DEFINITION OF A FEASIBILITY STUDY

This section clarifies what a feasibility study is by providing an extensive definition, explaining why studies are conducted, outlining their limitations, and defining the process of a feasibility analysis through the identification of four key factors.

What Is a Feasibility Study?

A feasibility study is an analytical tool used during a business development process to show how a business would operate under a set of assumptions. These assumptions often include such factors as the technology used (the facilities, equipment, production process, etc.), financing, (capital needs, volume, cost of goods, wages, etc.), marketing (prices, competition, etc.), and so on.

The study is usually the first time in a project development process that many key pieces and information about the project are assembled into one overall analysis. The study must show how well all of these pieces fit and perform together. The result will be an overall assessment of whether the proposed business concept is technically and economically feasible. Feasibility studies should also provide sensitivity analyses of the business given changes in key assumptions. One should note that a simulation or projection model, while useful, is not a substitute for a comprehensive feasibility study. This type of model is sometimes used in a "pre-feasibility" study done early in the project timeline to provide a first-cut evaluation of the proposed business idea.

The feasibility study evaluates the project's potential for success. The perceived objectivity of the evaluation is an important factor in the credibility placed on the study by potential members, lenders, and other interested parties. For this reason, it is important to hire a consultant with no formal ties to equipment manufacturers or marketers, for example, so that an unbiased evaluation of operating potential and efficiency can be made. Also, the creation of the study requires a strong background both in the financial and technical aspects of the project. For these reasons, outside consultants conduct most studies, although the project leadership normally has input as well.

Identify economic need

- Leaders and other potential member-users identify the economic need the cooperative might fulfill.
- A steering committee of potential member-users is selected to guide the project.

Deliberate

- Develop an overview of proposed business operations.
- Survey prospective members to determine the potential use of a cooperative.
- **Conduct a feasibility study.**
  - Economic
  - Marketing
  - Technical
  - Financial
  - Management
- Develop a business plan.

Implement

- Prepare legal papers and incorporate.
- Elect a board of directors.
- Implement the business plan.
- Conduct a membership drive.
- Acquire capital and develop a loan application package.

Execute

- Hire the manager.
- Acquire facilities.
- Begin operations.

Figure 1. The Cooperative Business Development Process.[1]

Feasibility studies for a cooperative are similar to those for other businesses, with one exception. Potential members use the feasibility study to evaluate how a cooperative business idea would enhance their personal businesses rather than to determine the return on investment they would receive on invested stock. A study conducted for an agricultural marketing cooperative, for example, must address the project's potential impact on members' farming operations in addition to analyzing economic performance

[1] See Appendix A and CIR 7 - "How to Start a Cooperative", for more information on these stages and the individual steps involved in the entire development process.

at the cooperative level. In other cases, such as food cooperatives, the value to the member is access to consumer goods or services, possibly at lower prices, and is not based on the economic return to the cooperative itself. Cooperative businesses are developed first and foremost to serve members' needs and enhance their economic well-being. However, to do so, they must operate efficiently and compete effectively in the marketplace.

Why Prepare Feasibility Studies?

Developing any new business venture is difficult. Taking a project from the initial idea through the operational stage is a complex and time-consuming effort. Most ideas, whether a potential cooperative or an investor-owned business, do not develop into an operational entity. When ideas do make it to the operational stage there is a high failure rate (many within the first 6 months). Thus, before potential members invest in a proposed business project, they must determine if it can be economically viable and then they must decide if investment advantages outweigh the risks involved—a feasibility study is the means by which these decisions are made. Without feasibility studies the percentage of startups that fail would be higher.

Many cooperative business development projects are fairly expensive undertakings that can also be confusing to potential members. Proposed cooperatives often involve operations that substantially differ from those of the members' individual businesses, and cooperative operations may involve risks with which the members are unfamiliar—another reason that a feasibility study is so important. It should provide a clear understanding of project risk to help members decide whether to invest in the proposed business.

Members participate in the development of the feasibility study and thus are educated about various aspects of the project, which will help them decide whether to move to the implementation stage. In addition, this knowledge helps prepare members of the steering committee to become the board of directors, as often happens if the project is implemented.

While the costs of conducting a study may seem high to the potential members, they are relatively minor when compared with the total project investment that will be required. The expenditure for a feasibility study is actually inconsequential if it saves an unprofitable venture from going forward, thus preventing the larger capital investment needed to start most new businesses, as well as the time and effort involved, from taking place. And if the study shows that a project is indeed feasible, it provides the group with some concrete useful data that can be used in subsequent business plans and projections.

Feasibility studies are useful and valid for many kinds of business development projects. Evaluation of whether to start a new business, by either new groups or established businesses, is the most common, but not the only usage. Studies can help groups decide to expand existing services, build or remodel facilities, change methods of operation, add new products, or even merge with another business. A feasibility study assists decision-makers whenever they need to consider alternative development opportunities.

Feasibility studies permit planners to outline their ideas on paper before implementing them. This can reveal errors in project design before implementation is made. Potential stumbling blocks can be identified and decisions made about whether they could be effectively addressed if the project goes forward. Applying the knowledge gained from a feasibility study can significantly lower overall project costs by keeping adverse designs and planning concepts from being made.

A feasibility study presents and clarifies the risks and returns associated with the project so that prospective members can evaluate them. There is no 'magic number' or correct rate of return a proposed cooperative needs to obtain before a group decides to proceed. The acceptable level of return and appropriate risk rate will vary for individual members depending on their respective personal situations and need for the proposed services of the cooperative.

Lender Considerations

A proposed project usually requires both risk capital from members and debt capital from banks and/or other financiers to become operational. Lenders typically require an objective evaluation of a project when they consider a loan investment, and a feasibility study often provides the first look at those aspects. While some groups often try to involve a lender early in the process, a feasibility study is often conducted with an eye toward explaining the project to potential financiers. Lenders have different requirements from the study than group members. Lenders are most interested in the project's ability to pay back loans while group members are interested in the benefits to them of using the cooperative.

Many groups work with lenders with whom they have an established personal or business relationship. This may expedite the process of obtaining financing. Nevertheless, the lender must know and understand the unique aspects of cooperatives and fully understand the characteristics and potential of a proposed project. The feasibility study will help them in this regard.

Lenders' primary concerns focus on repayment, their risk exposure, and a project's strengths and weaknesses. Lenders classify these concerns into the '5-C's':

- Capacity—what is the group's ability to repay the loan?
- Capital—what assets are being financed with the loan and how much is requested?
- Character—who are the principals of the project? What is their background?
- Collateral—what is being used to secure the loan? How is it valued?
- Conditions—what additional factors can affect the loan?

The odds for financing diminish if a lender does not fully understand the project and is unable to review the potential financial results through a sound economic and financial analysis. Success or failure of a business opportunity often hinges on obtaining adequate lender financing. For this reason, it is often a good strategy, if possible, to consult with potential lenders prior to conducting a feasibility study to determine what factors they will focus on given the type of project. Such a consultation can shorten the time that a lender needs to approve project financing, or even improve the ability of securing financing. However, while a feasibility study is important for providing information that will help in gaining finances from a lender, it should not be conducted merely to prove to them that a project is viable. It must only be undertaken when the proposed project is being seriously considered for implementation by dedicated potential members.

Feasibility Study Limitations

Although a feasibility study is a useful tool for project deliberation, it has limitations. A feasibility study is not an academic or research paper, but is a pragmatic information and data analysis document. It is confidential to the group for which it is conducted, and is not for public dissemination. A completed study should permit a group to make better decisions about the strategic issues of its specific project.

The study is also not a business plan, which is developed later in the project development process and functions as a blueprint for a group's business operations for implementation (see Appendix B). Given a group's decision to proceed after evaluating a feasibility study's results, the business plan presents the group's intended responses to the critical issues raised in the study. Many of the outcomes presented in the feasibility study form the basis for developing a business plan if a decision to proceed is made.

A feasibility study is not intended to identify new ideas or concepts for a project. These ideas should be clearly identified before a study is initiated. Assumptions that are partially developed from these ideas provide the basis for the feasibility study, so the more realistic they are, the more value the study's findings will have for a group's decision-making.

A study should not be conducted as a forum merely to support a desire that a project be successful. Rather, it should be an objective evaluation of a project's chance for success. Even studies with negative conclusions are useful for group decisions.

As stated earlier, financiers may require a feasibility study before providing loans, but this should not be a study's only purpose. Although a study can enhance a banker's ability to evaluate a project, the primary goal should be to aid a group's ultimate decision on going forward, not only whether financing can be secured.

A feasibility study will not determine if the project will be initiated, since that depends on the potential members, who will invest in and become the owners of the business. However, the information, data, and facts offered in a study, given realistic assumptions, provide the basis for a decision. Potential members must decide if the benefits justify the risks involved in their continuing the project and the study findings will assist them in that assessment. A study uses basic project assumptions to develop an analysis, shows how results vary when assumptions change, and provides guidance as to critical elements of a project. Conducting a study should provide the group with project-specific information to assist it in making decisions. This should lower the risk of continuing with a business development project that ultimately would fail.

First Steps in Feasibility Study Development

Some initial steps must take place as a group progresses toward fully assessing a project. First, the group must decide on whether to proceed with a feasibility study. Second, the project must be accurately defined. Third, there must be strong commitment and leadership from the group, and fourth, those involved must fully understand the process of making sound decisions.

Step 1—Decide Whether to Proceed with a Study

The first step for a group to take is to fully deliberate the necessity of even conducting a feasibility study. The group must carefully consider whether it is ready and prepared to have a study conducted. Because a group's resources are likely limited, it is important that the group be ready to proceed with the project and the feasibility study before allocating the necessary resources. It could very well be that the study needs to be put off until another time because of a lack of support from prospective members, not enough capital to proceed, or any other given reason. Once the decision is made to invest the time and resources in a feasibility study, the group proceeds to define the project.

Step 2—Define the Project

In a successful cooperative development project, a core group of people must feel a strong need to work together to solve a problem or take advantage of a business opportunity. Working together provides the context for a cooperative business project. What the project will entail must be understood and the group must believe the idea is worth pursuing. Often, a few individuals provide the spark for an idea, but group interaction permits them to hone an idea and develop sufficient interest. When defining a project early on, a group often discovers common interests that a potential cooperative business may be able to address. Cooperative businesses work best when participants see a mutual benefit from working jointly rather than acting alone to achieve a goal. Members voluntarily choose to belong to a cooperative because they see some potential benefit. When a project can be addressed jointly, potential member interest exists, and benefits are possible, then a cooperative can be the solution. To clearly define a development project, a number of factors should be included in spelling out the project idea. The project idea should be:

- Leaders and other potential member-users identify the economic need the cooperative might fulfill.
- Understandable (described in such a way that the objective is clear);
- Significant enough to warrant group action;
- Capable of providing economic and/or technical solutions to a problem or opportunity;
- Economically and socially fitting for a group; and
- Considered a reasonable business solution.

When all these elements exist in a project idea, the potential exists for developing a successful cooperative. If any are lacking, the concept should be rethought and the project definition revised before proceeding with the feasibility study steps.

A carefully defined project idea will provide the steering committee and group with a foundation from which to judge the project as it proceeds.

Here is an example of a project idea statement: ***A member-owned cooperative that will process and market members' soybeans for the farmers of "ABC valley," to meet the area's high demand for soybean meal and soybean byproducts, and to provide strong value-added economic benefits to members.***

The group then may provide some key points in addition to an idea statement to further clarify how the project will meet necessary economic, business, and technical factors.

Step 3—Group Commitment and Leadership

For a cooperative development project to have a chance at success, it must have individuals involved who are committed to the defined idea. A critical mass of potential members should be involved because having sufficient support for a proposed concept is crucial. The specific number for a critical mass will of course depend on the type of products involved, the scope of what the business will do, and the economic resources that will be needed. A smaller number of individuals who are fully committed to a project, provided they have sufficient product, capital, or demand for services, can have a higher chance of success than a larger number who are only partially committed. Thus, the key is not necessarily the number of people involved, it's the level of commitment they have. Although when a development project is a complex endeavor that will involve a highly technical processing plant (for example, an ethanol or biofuel processor, or a meat plant or grain mill), then having a larger number of committed people becomes very important.

Clearly, commitment and loyalty to an idea cannot be overstated. It can be measured several ways, such as attendance at organizational meetings, positive potential member survey results, the amount of willingness to become involved and personally invested in a project, and financial backing when requested. Financial support from potential members is perhaps the best measure of support. For instance, are people willing at the start to contribute to finance all or part of a feasibility study? Asking potential members for an initial contribution can help sort out those who are serious about the effort from those who are not.

The second step in the events for starting a cooperative outlined in Appendix A indicates that a steerng committee of potential members should be formed from a group for guiding the project. A steering committee represents the larger group and takes the major leadership role in the project.

Groups need people who are leaders to assume control of a project and to be part of the steering committee. An important prerequisite is that they be willing to join the proposed cooperative and commit to financing it and using its services. Chosen or volunteered leaders must be people that will be active participants in the development process as they will ultimately be responsible for making key decisions and plotting the project's direction within the steering committee.

The steering committee must also ensure that all potential members involved feel free to voice their opinions and viewpoints about a project. Different perspectives are important considerations for developing a new business, and if alternative strategies or options are suggested and deemed important to evaluate, they should be brought into play during the feasibility study process so they can be properly assessed.

In USDA's Cooperative Service Report 54, "Creating 'Co-op Fever': A Rural Developer's Guide to Creating Cooperatives" (see references or http://www.rurdev.usda.gov/rbs/pub/sr54/sr54.htm), author Bill Patrie mentions the following five characteristics of a "project champion" who provides strong leadership:

1) Credibility
2) Financial stability
3) Basic knowledge of the industry
4) Willingness to accept the servant leadership role
5) A developer, not a promoter

Patrie defines these five characteristics (verbatim) as follows:

1) Credibility—Is the individual personally credible in his/her neighborhood? They need not be the biggest farmer or the most active in commodity associations, but they must be respected for their judgment. Avoid individuals who have tried every new idea that has come around and are suckers for anything new. I look for people who finish what they start and can take a long-term view.

2) Financial Stability—Is the individual capable of keeping his/her house in order? Producers who have failed before (especially if they have gone through personal bankruptcy) usually lack the credibility with other producers and lenders to lead the project. They must be able to devote time away from their personal business to help develop the cooperative. This criterion is extremely limiting because many producers lack the time it takes to do the work without jeopardizing their individual operations. I once worked with a cooperative whose interim board chair wanted to use organizational funds to buy clothes. Her argument was that she would make a better impression on investors if she could afford to dress well.
3) Basic Knowledge of the Industry—Is the individual familiar with the industry in a comprehensive way? Most value-added cooperatives are also vertically integrated. The project champion must have a basic understanding of the entire industry—from the first steps of production through processing to marketing to the final consumer. This is a tall order and can't be easily filled. The "Madison Principles"[2] are critical at this stage of leadership selection.
Often, producers become enamored of a manufacturing technology or an available building and want to quickly close the deal to own the facility or the equipment. A true project champion must lead the group through a market analysis prior to analyzing processing facility and equipment needs. If an individual can't be found who has this basic understanding of the industry, then I look for a person who is willing to learn.
4) Willingness To Accept the Servant Leadership Role—The project champion is often uncompensated. They will frequently be criticized, often unfairly, and sometimes insulted. Thin-skinned or quick-tempered people often do not last in the pressure-cooker environment of creating a new cooperative enterprise. I look for a project champion who has balance in her/his life. They must have patience, people skills, a good sense of humor, and a sense of what is ridiculous.
5) A Developer, Not a Promoter—This is development work, not promotion. Promotion may get column inches in the local paper and a 30-second spot on the 6 o'clock news, but it won't build a financially

[2] The "Madison Principles" are 12 principles for cooperative development practitioners to follow. They were developed by the members of Cooperation Works! in Madison, Wisconsin, in 1985. See http://www.cooperationworks.coop/about/madison-principles.

viable company. While enthusiasm is important, it can't replace critical common sense and solid business judgment.

These five attributes are important in a "project champion" or leader of a cooperative development project.

Use of Advisors and Consultants

Outside advisors and consultants can be useful to a group during the business formation process. However, outsiders, no matter how well intentioned, should not be put into overall leadership positions. If they are, the process often becomes more "top down" directed rather than internally directed by prospective members, and in this case, potential conflicts may arise and the focus of the group's vision may be skewed.

At the same time, a group should feel free to seek outside experienced consultants to help guide it through the development process, or perhaps to aid just a specific aspect of the process. For example, extension agents or lenders that interact closely with a group may be willing to help; accountants and lawyers may provide assistance in specific areas such as bookkeeping, legal structure, and drawing up legal documents; and advisors, such as USDA cooperative development specialists or development practitioners from a cooperative development center, can help the group with all or some aspects of the development process, and may even provide direct technical assistance with feasibility studies or business plans.

Outside consultants are useful because of their experience and expertise with the development process and because they also work in an objective manner to ensure that all potential members' ideas, thoughts, and concerns are considered and that assumptions and information are accurate and realistic. (See the "Feasibility Study Key Actions" chapter for more information on choosing a consultant.)

Step 4—Understand Sound Group Decision-making

As stated above, strong and committed leaders are essential for defining a project and deciding if a feasibility study should be conducted. Informed leadership with enlightened self-interest and a commitment to group action is needed. Leaders must maintain a strong focus on the decisions to be made and be able to create an environment where participants are encouraged to be active and involved in discussions, creativity, and decision-making.

For a project to succeed, all potential members of a new cooperative business venture must be kept informed about project details as they evolve so that they buy into and feel committed to the project.

To assist with decision-making, those on a steering committee should ask two questions: (1) If a bad decision is made, what would be the cost? and (2) If no decision is made, what would be the cost? If the cost of making a wrong decision is relatively small, do not spend much time, money, or effort on the decision-making process. On the other hand, if the cost of committing an error could be large, it's better to put more resources into determining the pros and cons of the decision and defining all of the issues before choosing an option.

In practice, this is not always easy to implement given the different personalities involved and the personal preferences for making decisions. Some may be slower to learn, need more time to contemplate before making a decision, or have aversions to high risk. On the other hand, some members may want to go ahead and make a decision before relevant information has been gathered and fully assessed. Balancing diverse aspects within a group can be difficult, but working to do so is paramount.

Thus, decision-making often is one of the greatest initial challenges that a group faces in developing a project. Figure 2 presents some guidelines to assist groups with the decision-making process. Given the difficulty in making decisions, some groups or individuals try to avoid it. There is always more information that can be gathered, but there is also a cost to taking more time to deliberate. A decision must be made when further investigation costs more than new information is worth.

- Unanimous agreement is not required to move forward; a consensus approach is better.
- Never decide to proceed based solely on negative reactions, such as resentment or envy toward middlemen, lenders, etc.
- A few reliable persons are superior to a larger number of doubtful persons.
- Base decision-making on economic and social realities faced by the cooperative.
- Make each decision only once.

Figure 2. Guidelines for Group Decisions.

FEASIBILITY STUDY KEY ACTIONS

Once a steering committee and group have made the decision to proceed with a feasibility study, there are a number of key actions that need to be taken. Figure 3 provides, in chronological order, the actions or decisions that have to be made. The key actions for a feasibility study include: deciding who will conduct the study, development of project assumptions, determining which components (study areas) will be included to make the study comprehensive, accepting or rejecting the completed study, and group decisions after accepting the study.

1. Deciding who will conduct the study.
2. Development of project assumptions.
3. Determining components for a comprehensive study.
4. Accepting/rejecting the study.
5. Group decisions after accepting the study.

Figure 3. Key Actions for Feasibility Studies.

1. Deciding Who Will Conduct the Study (Consultant Selection Criteria)

Although in principle it is possible for a project group member to conduct the feasibility study, normally, an outside consultant is hired to do it. Most prospective members and financiers view an objective evaluation of a project concept via an outside practitioner as important. This objectivity often provides a group with helpful information that might have been overlooked by one who is participating directly in the project.

Hiring a consultant to create a feasibility study is an important decision, and thus the steering committee or group must use care when selecting that person (or firm). In practice, consultants have differing levels of ability and usually a consultant will be strong on some points and weaker on others. The key is to select a feasibility practitioner who is skilled in cooperative development and versed in areas relevant to the type of project.

Figure 4 provides possible criteria to use for selecting a qualified consultant. The steering committee will need to determine if a consultant is technically proficient enough to undertake a feasibility study and whether he or she has significant experience in doing so. The committee should review samples of previously prepared studies and speak with others for whom the person or firm has worked before contracting with them. It is important that a consultant have the traits required to work well within group situations.

- Has previous experience conducting feasibility studies.
- Has experience with the industry to be studied, or access to experience and associated professionals.
- Works independently and objectively (e.g., of equipment manufacturers, marketers, etc.).
- Understands cooperatives fully (their operations, governance, financial workings, etc.).
- Is willing to listen to the groups' ideas.
- Works closely with designated contact members of the steering committee or group.
- Is willing to revise study given feedback.
- Accomplishes the study within an agreed upon timeline.
- Works within the group's designated budget.
- Is a strong writer with skills in data analysis and spreadsheet design and presentation.
- Provides clear, useful information in the completed study.

Figure 4. Criteria of a Good Feasibility Study Consultant.

Consultants should have experience in the industry under study. Otherwise they may not correctly identify critical factors. Given business complexity, it is almost impossible for one person to have experience in all areas. Some consulting firms resolve this issue by having their feasibility specialist work with contracted industry experts. In any case, it is important to research many sources for all the pertinent information possible about an industry.

A team approach may, in some instances, be utilized to develop a study. For example, a cooperative development specialist from the USDA or a practitioner from a cooperative development center could work jointly with industry specialists to create a feasibility study.

The consultant should also understand the unique aspects of cooperatives. Tax implications, distribution of net margins (profits), management, and other business considerations (e.g., governance) of cooperatives differ from those of other businesses and the nuances of each must be properly presented.

The consultant should avoid preconceived notions about how the project will function. The study should not be an "off-the-shelf" document assembled from previously created studies. Rather, the consultant should pay particular attention to the ideas that the group has developed and craft a unique study suited to the group's needs. The consultant should work closely with the group and be receptive to its suggestions. Also, the consultant should be prepared to make technical revisions or to correct errors given group recommendations and wishes. Revisions are a normal part of the study development process. Revisions should focus on the validity of the assumptions and the technical design of the study.

Using an outside consultant brings objectivity to the feasibility study rather than merely providing the results that the group wants. Consultants have a legal obligation to provide a responsible analysis. They should not be asked to alter the results merely to conform to members' desires for a project's viability.

Timeliness is an important consideration when selecting a consultant. Projects are time sensitive. Usually, decisions to proceed await information provided in the feasibility study. So care and diligence required for a well-crafted study must be balanced against the desire for speed. A qualified consultant must be able to complete a well-designed study within a timeframe that serves the group's needs. On the other hand, the timeline must be realistic. And, a consultant can only progress as fast as a group makes the required decisions, provides information to the consultant, and carries out its other project responsibilities.

Cost is an important factor. The expertise and skills that consultants offer a project must be weighed against their cost. A quicker timeline could increase a consultant's fee. Preparing a pre-feasibility analysis may decrease the effort required to complete the feasibility study and reduce the cost.

Some public programs offered by the USDA's Rural Business-Cooperative Service, community development offices, the Small Business Administration, some cooperative development centers, and local business incubator programs provide technical assistance at little or no cost to groups creating feasibility studies. There are also grant programs available such as USDA's Value-Added Producer Grants program, which can provide funding for a feasibility study if a project meets the program's criteria and is selected. This program requires a one-to-one matching contribution from the applicant.

A consultant should provide the data used to generate the financial tables and scenarios reported in the feasibility study and, preferably, an electronic spreadsheet format that can be easily manipulated. Although requesting this information can moderately increase the cost of a feasibility study, access to the actual data permits the group to use the information for later needs with greater flexibility. This data can also reduce the cost of creating the business plan, if the group proceeds to that stage. Additionally, it can decrease the effort required for revisions, if in the future the group changes the project's assumptions to differ from those in the study.

Once the consultant has been selected, the group should provide detailed instructions on the study requirements. There should be a legally binding contract between the parties. The group should consult legal counsel for assistance. The contract should state clearly the requirements and role of both the group and the consultant. It should have timelines, delivery dates, explicit deliverables, and what is to be accomplished before payment is made. Often, the consultant receives a downpayment before the feasibility study has been conducted. The balance is paid only after the study has been reviewed and accepted by the group (and possible financiers if appropriate). This gives the group more leverage to encourage timeliness or revisions. The contract should designate a third-party arbitrator to resolve any disputed items.

A complex, large-scale project may require several consultants to complete various aspects of the study. Multiple consultants can reduce the group's dependency on a single person or company. It also can permit the group to select experts from several fields. However, it also can complicate the coordination and consistency of the information received.

Before signing the contract, the group should discuss with the consultant arrangements for cost overruns, time delays, revisions, and what considerations will be made for these issues. Changes after signing the contract can be costly or delay the study results. All parties should be clear about what to expect prior to signing the contract and initiating the study. (See Appendix C for a point-award system for selecting a consultant based on select criteria.)

Feasibility Study Working Relationships

A few qualified members of the steering committee (if the committee is a large one), or the entire steering committee (if it is a small one) should be designated to work closely with the consultant or person developing the study. These group members must see that the feasibility study properly presents and reflects the right aspects of the project as it has been designed, and in accordance to the defined assumptions. Through this working relationship the study should be tracked through all of its stages and its ideas reviewed and clarified.

Steering committee members with appropriate backgrounds and the ability to commit sufficient time to working with the consultant should be selected. These contact members represent the group's interests to the consultant. They are the contact to provide clarification and additional information that the consultant may require. Plus, they should provide periodic reports to the group about the study's progress. They also should work with other group members and advisors to gather the information needed for the feasibility study. These members are obliged to express the wishes of the entire group and not just their own views.

Members or outside financiers will often perceive the reliability of the entire study based on its least accurate piece. An otherwise well-conducted feasibility study could be viewed as inaccurate or useless because of a simple mistake. To prevent this, the feasibility study should be carefully examined for overall clarity and logical consistency—is the language appropriate; is the document well organized; and can someone who is not familiar with the project understand the study and its findings? Reviewers should confirm that the study's assumptions are clearly documented, well described, justified, and as accurate as possible.

Although the contact members take the lead in working with the consultant, others should review the study carefully before the group decides to accept it. Advisors such as USDA cooperative development specialists or Extension agents can provide an objective review and offer insights on content or study assumptions. This outside review can be especially useful when the group has used consultants to prepare the report. Often, a series of draft reports are presented to the group as the study proceeds. Issues identified that warrant changes to the study are then conveyed to the consultant.

2. Development of Project Assumptions

Key project assumptions should be determined at the initiation of a feasibility study. Usually an assumption is thought of as something that is taken for granted, but in the current context assumptions provide the basis for the project and therefore need to be carefully thought out and developed. Since the group cannot analyze every variation of a project, it must provide boundaries within which the study will be carried out. The consultant, if one is being used, can assist in the development of assumptions by providing objective knowledge and expertise. He/she should also ask the group difficult questions to narrow the range of assumptions and make sure they are as accurate as possible, as well as justifiable.

Figure 5 provides four questions and some clarifying statements that a steering committee or group should address as it develops assumptions for the feasibility study.

The steering committee or group may not be able to provide all of the required detail for each of the assumptions that need to be developed, so again, using an experienced practitioner/consultant to help research and develop some of the assumptions may be necessary. Furthermore, some of the assumptions will have more than one option to study. That's where sensitivity analysis will come into play. Other questions that help determine proper assumptions might arise as well, depending on the type of project. It is up to the steering committee and those conducting the study to explore all avenues when determining the assumptions needed for a full analysis in a study.

Considering more than one potential business structure and/or alternative business process is not a problem at this stage. However, it is important that an analysis be conducted in the feasibility study for each identified project scenario so that the steering committee or group can assess them.

3. Determining Components of the Feasibility Study Report

A comprehensive feasibility study will contain all the ingredients necessary for the steering committee and group to make a sound decision on whether to proceed with a project. Although studies vary depending on the type and scope of the proposed business, all reports must contain enough elements to present a comprehensive view of the project. While some specific project details may be undecided, such as plant location or who the manager will be, a report must contain enough information and analyses to determine a project's potential for success or failure.

The feasibility study report serves as the written representation of the group and its potential cooperative business. Potential members, financiers, and others will use this document to help determine their level of support for the project. The report's appearance as well as its content can influence people's perception of it. Thus, the layout should be professional, well organized, and well written.

The appearance of and specific aspects included in the report will vary depending on the project, the group, and the consultant who prepares the study. Thus, there is no required length or number of components for a study report, but the study must provide an organized format with enough critical information and analyses pertinent to the project to help the group make an insightful decision.[2]

Key elements will change depending on the nature of each project. As a rule of thumb, if reasonable changes in a factor could make the project change from successful to unsuccessful, it is a key element.

1. How/why is the proposed cooperative needed (as determined by the potential members)?

- Define the assumed products and/or services to be handled or provided (there can be more than one and each should be clearly defined).
- Explain the proposed cooperative's comparative advantage (e.g., define what the market is demanding and what producers do well).
- Describe the proposed cooperative's benefit to members (e.g., enhanced marketing, higher marketing prices, lower prices for products purchased, more efficient and lower cost services, etc.).

2. What is the potential membership base and volume of product for the project? (This data is normally gathered via a survey of potential members.)

- Define the level of potential support by producers who would have the opportunity to participate.
- Describe the approximate number and size of the producers who are likely willing to participate.
- Define the potential volume of products or services.
- Explain the potential for future expansion of membership and volume.

3. How well will the cooperative fit into the market?

- Define the projected prices for both inputs and outputs.
- Define the projected volume of sales.
- Explain the size of the market and how the cooperative fits in (e.g., market share).
- Determine the potential for strategic alliances.

4. What are the financial and organizational needs for the project?

- Estimate overall capital needs and describe potential sources of this capital.
- Define the level of financing needed and potential lenders.
- Describe the legal requirements, documents or agreements, permits, and inspections.
- Describe the facilities and equipment needed and whether they will be purchased, built, or leased, and estimate how much they will cost.
- Estimate the management requirements and skills, and the cost of obtaining the appropriate management.

Figure 5. Questions for Developing Assumptions.

Examples could be the technology of production, volume of inputs, the market for goods sold, marketing channel, personnel costs, prices paid, and capital costs. Figure 6 provides a general example outline of the major components a feasibility study might contain. This example includes eight major components, but the exact number and order of components for different studies could very well vary from these. In addition to the potential items listed in the outline below or others determined given the project, a study should include a title page, the name of the person(s)/firm who conducted the study, and a table of contents. General descriptions of each of the sample components included in Figure 6 are described in the following sections (sections include relevant outline items).

Executive Summary

I. Executive Summary

A. Summary of the Important Findings and Recommendations

It is important to have a concise summary of the critical segments of the report in an executive summary at the front of the report. This will allow reviewers to gain a strong sense of the report's significant information and major findings before they proceed with reading the entire study. Each major part of the report should be briefly and clearly summarized. When applicable to the major findings and final conclusions, significant data reflecting concrete analysis should be provided, and a summary of the key recommendations listed. This segment of the report should provide a context from which the reader will be able to better decipher all the components and findings of the report. As a means of setting the foundation for the study, it is important to also identify the steps completed for the project up to the current point in time, and the names of those heavily involved (the steering committee members at least).

I. Executive Summary
- A. Summary of the Important Findings and Recommendations

II. Introduction—Project Description and Justification
- A. Description of the project
- B. General setting and need for project
- C. Work already completed, pertinent dates, and those involved in the project

III. Industry Background
- A. Basic background information on the industry
- B. Economic conditions of the industry
- C. Implications and feasibility of entering industry

IV. Marketing
- A. Market potential for goods or services to be handled
- B. Markets to be served (current and future) and their attributes
- C. Ease or limitations of entering the market
- D. Marketing plan (strategies to be followed, summary of key actions)
- E. Overall assessment of the marketing situation and plan

V. Operational and Technical Characteristics
- A. Supply of labor and its quality (including management)
- B. Supply of key inputs needed for operations
- B. Technical characteristics and specifications of required plant and equipment
- C. Assessment of potential operational capacity and efficiency
- D. Location considerations (if one has not been already selected) and assessment (if one has been selected)

VI. Financial Statements and Projections (pro forma statements)
- A. Projected revenues, operating costs, and net income
- B. Capital requirements, potential and actual sources of equity, accumulation schedule, investment schedule (plant, equipment, human resources, etc.)
- C. Pro forma cash flow statement
- D. Income, balance sheet, and sources and uses of funds statements

E. Equity accumulation plan and financial ratio analysis
F. Financial plan summary (description of how it will all fit together)

VII. Summary and Recommendations
A. Concise Summarization of the Major Findings
B. Recommendations and Concluding Comments
C. Development schedule (remaining key steps and accompanying dates for action)

VIII. Appendix
A. Appendices (additional spreadsheets)
B. Important supplemental information
C. Notes, credentials, and references

Figure 6. Example Outline of Feasibility Study Report Components.

Introduction

II. Introduction—Project Description and Justification

A. Description of the project
B. General setting and need for project
C. Work already completed, pertinent dates, and those involved in the project

This section is usually somewhat brief and simply introduces the cooperative project and provides some justification for its need. Information as to when the project process began and in what stage it is now should be offered. In general, size and scope of the project, membership aspects, methodology employed for data collection, marketing and economic conditions, competition, relevant technical factors, economic and community conditions, etc., can all be briefly introduced to provide the reviewer/reader with an overall general conception of what the cooperative project entails.

Industry Background

III. Industry Background

A. Basic background information on the industry
B. Economic conditions of the industry
C. Implications and feasibility of entering industry

The state and status of the industry within which the cooperative will operate should be described in as much detail as possible and be broken down into geographic applicability (i.e., foreign, domestic, regional, local) to the project. The study should include charts and graphs of industry trends (e.g., volume, prices, byproducts, etc.), as well as a complete assessment of the competitive environment to properly define the need or fit of the cooperative within the industry sector. Pertinent data from industry organizations is helpful if it can be acquired. Government regulations and policies within the industry in question should be fleshed out and their relevance to the proposed business explained. Any regulations that might need to be met (e.g., environmental impact assessments, permits, etc.) should be clarified and analyzed. Costs associated with the government regulations and policies of an industry will need to be documented for the financial projections section.

Marketing

IV. Marketing

A. Market potential for goods to be handled or services to be provided
B. Markets to be served (current and future) and their attributes
C. Ease or limitations of entering the market
D. Marketing plan (strategies to be followed, associated costs, summary of key actions)
E. Overall assessment of the marketing situation and plan

Various components of the project's proposed marketing plan, whether for products to be marketed or goods to be sold, need to described and analyzed. The marketing environment should be fully described. The description should include how the product or products will be introduced and channeled into available markets. A description of potential customers, processors, handlers, etc. should also be provided. Procurement and sales strategies for commodities or goods to be purchased and/or sold should be described. This section should address market demand implications, marketing costs, transportation issues, coordination with others in the market chain (e.g., brokers, venders, manufacturers, processors, pooling, etc.), the quality and form of the products to be marketed, and an overall strategic assessment of marketing the product or products. When applicable, information from market outlook reports (e.g., USDA and other government agencies) that provide forecasts on specific crops, products, and industries are helpful for providing a context for the marketing plan.

Relevant charts, graphs, and tables should be provided to present a clear picture of the marketing environment. If it's a value-added venture, the implications of marketing the resulting products should be defined. How those products fit into existing markets given competitors' similar products should be researched and reported. From the overall marketing analysis, an assessment of the feasibility of the proposed marketing plan should be included.

Operational and Technical Characteristics

V. Operational and Technical Characteristics

A. Supply of labor and its quality (including management)
B. Supply and costs of key inputs needed for operations
C. Technical characteristics and specifications of required plant and equipment
D. Assessment of potential operational capacity and efficiency
E. Location considerations (if one has not been already selected) and assessment (if one has been selected)

This section lays out the operational aspects and procedures of the proposed business including: the supply of labor and its quality; which key inputs will be required (raw materials such as soybeans or wheat, for example), their source (supported by a survey of potential members, if applicable) and their cost; the technical characteristics (e.g., the type of plant design required, equipment, facilities, building systems, etc.); the feasibility of finding proper management; location aspects; and operational issues or options; etc. The study should address the ability of the project to operate efficiently within the scope of the project's parameters. It is important to provide information on the technical aspects of the project and to show how the proposed technologies will work within the context of the entire project. In projects with unproven technologies, this can be the most important aspect of a study and it provides a basis for close assessment. In projects with proven technologies, the study can serve to correct design flaws before costly mistakes are implemented. If the project requires construction of a sophisticated facility, such as a meatpacking or soybean processing plant, professionals such as architectural, engineering, or management specialists will need to be consulted early in the process. The needed expertise should be described in the feasibility study. Assistance that will be needed for loan agreements, legal contracts, and construction should be documented also.

If a location has been selected, the study should address the implications of that location—is it efficiently situated for the potential labor supply, is it adequate for delivery and distribution channels, does it meet city/town ordinances and regulations, will permits be required, resources be available to cover its costs, etc.? If a location has not been selected, the feasibility study may provide some prerequisite stipulations, data, and standards by which to choose a location given the type of project, industry, and technology involved.

Financial Statements and Projections

VI. Financial Statements and Projections (Pro Forma Statements)

A. Projected revenues, operating costs, and net income
B. Capital requirements, potential and actual sources of equity, equity accumulation schedule, investment schedule (land, plant, equipment, human resources, etc.)
C. Pro forma cash flow statement
D. Income, balance sheet, and sources and uses of funds statements
E. Equity accumulation plan and financial ratio analysis
F. Financial plan summary (description of how it will all fit together)

Possible economic outcomes are a prominent part of a feasibility study and are critical in the overall assessment of a project. Therefore, it is extremely important to do a thorough and careful job with the financials. Financial projections are usually made for 3 years. Cash flow statements should be monthly, while income statements and balance sheets should be monthly or quarterly for the first year and then annual for the second and third years. Financial statements and projections stem from valid and objective assumptions. Financial assumptions, such as capital requirements, equity needs, prices, human resources needed, and other factors, will come into play here. Because the economics of the project are so important to project assessment, assumptions must be in line with the reality of the situation and should not be overly optimistic or simplistic. Assumptions such as price forecasts/projections should be based on solid facts, such as historical prices and changes that have occurred in the industry which may affect the outlook. The sources for the facts and the rationale for key assumptions should be well documented either in the report body or in an appendix. Most feasibility studies begin with pro forma cash flow statements based on the assumptions and other data collected about the project, such as equity collected, product volume, purchases, sales, and expenses, for example. Besides equity, revenue streams and operating costs, the pro forma statements must include repayment and interest on potential short-term and long-term debt and/or other investments in the project. The cash flow statements (usually done on a monthly basis) must clearly show when capital is introduced and when it is repaid. This is important for indicating the project's repayment capacity, a critical consideration for a lender or investor. For a sample pro forma cash flow statement, see Appendix E. Also included in this section are income statements, balance sheets, and sources and uses of funds statements (or statements of cash flows). These pro forma statements provide important information beyond the cash flow analysis. The plan for accumulating needed member equity adds even more information by providing dates, sources, and amounts of equity expected (this information will be likely obtained from a potential member survey). Another useful analysis to include is a ratio analysis where ratios are developed from the pro forma statements. For example, current ratios, debt ratios, assets turnover, return on net worth, return on investment, return on sales, etc., should be formulated and compared during the projected years. For sample pro forma operating, balance sheet, and ratio statements, see Appendices F, G, and H, respectively.

In the financial analysis, the study should show the impact of varying key project assumptions. This controlled variation, called sensitivity analysis, permits planners to view which project elements are the most susceptible to positive and negative changes. For example, what impact does a 10-percent reduction in sales volume have on net margins?

The sensitivity analyses conducted should then be studied, and those that are potentially realistic should be developed into specific scenarios, which would involve looking at all aspects of how the proposed possible changes would affect the project. Both "worst-case" possibilities and optimistic scenarios should be created for comparison purposes. A comparison table and discussion should be developed so that it's easy to assess the differences between scenarios. The financial section should summarize all the findings of the financial analyses and provide an overall assessment of the financial and economic implications of the project. The financial impacts at both the cooperative and member level should be detailed.

Summary and Recommendations

VII. Summary and Recommendations

A. Concise Summarization of the Major Findings
B. Recommendations and Concluding Comments

C. Development schedule (remaining key
D. steps and accompanying dates for action)

To finalize the study, the last section of text should summarize all of the major points that the information and analyses throughout the report provided. This will allow the reader to fully comprehend all the different pieces of the study and how they work in conjunction with each other. The project's impact on potential members should be addressed. Project benefits, for example projected payment to members for product delivered to the cooperative, and patronage refunds should be summarized. Potential members should gain an understanding of the benefits the proposed cooperative would provide them and be able to use that information to decide whether to join. The section should clearly describe any important factors that the steering committee needs to consider as it works toward implementing a full assessment of the business and making a decision on whether to proceed. Some other aspects of the study that might be covered in the summary include any possible project risks for potential members or other investors, potential legal and governmental setbacks that could come into play, and time-critical factors, among others. If the study shows that the project is clearly feasible, this section should describe any important work that still needs to be done and actions that need to be taken (with relevant dates) as the group works toward a solid business plan and implementation. If the study found that more information, resources, etc. are needed before the project will be feasible, it should clearly state such discrepancies and provide recommendations for potential actions that could alleviate the issues.

Appendix

VIII. Appendix

A. Appendices (additional spreadsheets)
B. Important supplemental information
C. Notes, credentials, and references

The appendix of the report should include supplemental tables, spreadsheets, charts, and information—that are related to the analysis and descriptions in the report's body—that will provide the reviewer with a greater understanding of the project. Some examples of supplemental information, which will be highly dependent on the type of business being studied, might include:

- Background information on assumptions used in the analysis if not fully described in the body of the study (some might be derived from potential spreadsheet data given as examples here).
- Monthly inventory tracking spreadsheets for commodities to be handled, purchased,processed, sold etc.
- Monthly sales price spreadsheets for commodities to be handled, purchased, processed, sold, etc.
- Capital purchase and depreciation schedules for land, buildings, equipment, parts, etc.
- Employee schedules and salary/wage information for any staff that will be hired (management, sales representatives, administrative staff, warehouse personnel, laborers, etc.).
- Debt repayment schedules for different categories of borrowing (real estate, equipment, working capital, etc.).
- Pro forma financial statements (cash flow, operating, balance sheet, etc,) for different scenarios studied, but that weren't a major focus in the body of the report.
- Other industry or territorial information, such as commodity or product alternative uses and sources, commodity processing yield data, demographic data, competitive data and mapping, etc.
- Credentials of those involved in developing or assisting with the study.
- References used in the study and resources that will be useful as the project progresses.

In some cases, this stated supplemental information will have been addressed in the study's main sections, so it won't have to be included in the appendix unless more information is deemed to be required.

4. Accepting/Rejecting the Study

The steering committee usually makes the preliminary decision to accept or reject a completed feasibility study. The steering committee—which has been working closely with the consultant—has the most knowledge of the feasibility study and thus should make a recommendation to accept or reject the study. The final decision then rests with the entire group after a full discussion. The decision to accept or reject a consultant's work should not be influenced by the findings of the feasibility study, but rather by its quality. A well-crafted, but negative, study can prevent learning the same information later in the project process at considerable trouble and expense. By the same token, a feasibility study with positive returns should not be accepted merely because it makes the project seem possible. Thus, the primary objective of the study is not to promote the business start-up but rather to provide an honest evaluation of the project's feasibility; that is, its prospect for success. Study approval should be based on the study's technical merits. Does it fulfill the work expectations that the group had when contracting with the consultant? Are the study assumptions reasonable and well explained? Is the project conceptualized in a manner very similar to what the steering committee communicated? Does the study contain significant facts, analysis, and accuracy? Is the study sufficiently comprehensive for a full analysis of the project? If key information is lacking or not felt to be properly analyzed, the study should be revised. If the committee thinks that other marketing avenues should be explored, or that changing conditions warrant further study, for example, then it should ask for those analyses to be done. In most cases, if major changes occur to the project idea as presented in the feasibility study, the group should have the consultant revise it to reflect these changes or initiate a new study. This permits the group members to make decisions with all applicable information.

5. Group Decisions after Accepting the Study

After the study's quality has been deemed acceptable, the steering committee and group need to decide whether to proceed with the project. Positive results from a feasibility study do not necessarily imply that the group should proceed with the project. Several factors could cause the group to stop or to revise the project:

- The situation/environment has significantly changed since work on the study was completed;
- The group has chosen another project it considered more beneficial;
- The risks are deemed greater than the group is willing to accept;
- Capital, size, or capacity requirements are more than the group can accommodate; or,
- New information shows key study assumptions to be unrealistic.

Negative study results do not necessarily signify that a group should stop developing the project. The group may cautiously proceed even if study results are negative. Any decision to continue should carefully weigh the risks involved and openly declare those to all involved before making a decision to proceed. Here are some reasons for the group to consider continuing with a business plan and project implementation when the study did not provide favorable results:

- The situation/environment has improved since the study was completed;
- Critical assumptions of the study are found to be unduly harsh or negative, or have significantly changed;
- More potential members and/or product volume have been identified;
- The group feels that more producers or volume will participate once the project is closer to implementation;
- The group has found a partner to share the cost, risk, capacity, etc; or,
- Technical limitations of machinery or design have been resolved.

The group should not proceed to develop a business plan with negative issues still pending. It is important that the steering committee and group address any recommendations and limitations the feasibility study outlines before it takes the time and approves the expense that a business plan will take. If a decision is made to proceed with the project, the steering committee and group should first look at the study's recommendations to see what, if anything, needs to be accomplished before a business plan is developed. For example, does the study advise exploring joint ventures with processors or other industry partners or organizational structures (such as a limited liability company), obtaining marketing contracts from prospective members, getting attorney assistance to meet federal or state security laws, researching other marketing avenues, etc.? Written records of the decision-making process should be made and retained. The steering committee and group have a legal responsibility for adequate due diligence. An attorney should be apprised of project developments as they occur—in this case the acceptance, rejection, or need for further analysis—of the feasibility study. The attorney needs this information to provide appropriate legal counsel to the steering committee and group as it proceeds. If all issues, recommendations, and limitations are fully explored, and the project is declared feasible, the group and steering committee proceed to develop a business plan (which is part of Step 7, in "How to Start a Cooperative", CIR 7). Many components and analyses contained in the feasibility study will be used in the business plan. The steering committee and consultant should work to identify those parts that are relevant and acceptable for inclusion in the business plan. With the development of the business plan, the steering committee and group will work toward completing the remaining events/steps of development, as explained in CIR 7.

References and Information Sources

Patrie, William. "Creating 'Co-op Fever': A Rural Developer's Guide to Forming Cooperatives," Service Report 54, Rural Development, United States Department of Agriculture, Washington, DC, July 1998.

Rapp, Galen and Gerald Ely, "How to Start a Cooperative," Cooperative Information Report 7, Rural Development, U.S. Department of Agriculture, Washington, DC, July 1998, Reprinted August 2001.

"Co-ops 101, An Introduction to Cooperatives" (CIR 55), a good information source for those needing a greater understanding of cooperatives, is available from USDA's Cooperative Programs (http://www.rurdev.usda.gov/rbs/pub/cir55/cir55rpt.htm. (Reference follows)

Frederick, Donald, A. "Co-ops 101, An Introduction to Cooperatives," Cooperative Information Report 55, Rural Development, U.S. Department of Agriculture, Washington, DC, April 1997, revised March 2005.

The Agricultural Marketing Resource Center has extensive information on value-added businesses available at www.agmrc.org. This site also provides links to resources for feasibility studies and business planning. Outside advisors can assist in the development process as well as be providers of other sources of background information. There are a number of cooperative development centers around the nation with the sole purpose of being practitioners of cooperative development. Some of these centers belong to the "Cooperation Works!" cooperative development network (http://www.cooperationworks.coop). However, there are centers outside of that network as well. Those interested in developing cooperatives can benefit from contacting a cooperative development center in their state or region (if there is one) for assistance.

USDA Websites:

U.S. Department of Agriculture: http://www.usda.gov/USDA Cooperative Programs: http://www.rurdev.usda.gov/rbs/coops/csdir.htm

USDA Cooperative Programs (publications):
http://www.rurdev.usda.gov/rbs/pub/NEWPUB.htm
USDA Rural Development: http://www.rurdev.usda.gov/

USDA Rural Development can also assist you from its State Offices. Please look up and contact the Rural Development State Office within your state: http://www.rurdev.usda.gov/rbs/coops/cscontac.htm

Outside websites/resources:
University of Wisconsin Center for Cooperatives:
http://www.uwcc.wisc.edu/ Quentin Burdick Center for Cooperatives:
http://www.ag.ndsu.nodak.edu/qbcc/

APPENDIX A. SEQUENCE OF EVENTS IN COOPERATIVE DEVELOPMENT[3]

1) Invite leading potential member-users to meet and discuss issues. Identify the economic need a cooperative might fill.
2) Conduct an exploratory meeting with potential member-users. If the group votes to continue, select a steering committee.
3) Survey prospective members to determine the potential use of a cooperative.
4) Discuss survey results at a second general meeting of all potential members and vote on whether to proceed.
5) Conduct a needs or use cost analysis.
6) Discuss results of the cost analysis at a third general meeting. Vote by secret ballot on whether to proceed.
7) Conduct a feasibility analysis and develop a business plan.
8) Present results of the feasibility analysis at the fourth general meeting. If participants agree to proceed, decide whether to keep or change the steering committee members.
9) Prepare legal papers and incorporate.
10) Call a meeting of charter members and all potential members to review and adopt the proposed bylaws. Elect a board of directors.
11) Convene the first meeting of the board and elect officers. Assign responsibilities to implement the business plan.
12) Conduct a membership drive.
13) Acquire capital and develop a loan application package.

[3] From "How to Start a Cooperative", Cooperative Information Report 7 (page 4, Figure 1).

14) Hire the manager.
15) Acquire facilities.
16) Begin operations.

APPENDIX B. THE FEASIBILITY STUDY VS. THE BUSINESS PLAN

Groups sometimes confuse the role of two tools used in business project development—the feasibility study and the business plan. The feasibility study helps determine whether to proceed with implementing the business while the business plan spells out how it will be implemented. Each has common components. Assuming positive feasibility study results, much of its information is incorporated into the business plan. The feasibility study is conducted during the deliberation phase of project development before financing is secured. It shows if the project concept can be viable.

This analytical tool includes several scenarios for the group to use in determining if it continues the project. If, after completing a feasibility study, the group decides to not proceed, there is no need to create a business plan. If the group decides to proceed, it prepares a business plan for project implementation. The plan serves as a blueprint not only for implementation but also for what actions the group will take during project operations.

The business plan usually contains less emphasis on scenarios than the feasibility study. Typically, it highlights only the scenario selected by the group as the most promising. The business plan is much more focused on what action steps will be taken during and after project implementation. The business plan is created after the feasibility study. Project details, which required assumptions for the feasibility study, have been decided. Standard business plans include details such as key management personnel, business location, the financial package, product flow, and possible customers. The feasibility study should be an independent review of the project by one or more experts outside of the group.

In contrast, the group itself typically develops its business plan internally, sometimes with the assistance of a consultant. It needs to be based on group members' vision for the business, since they will be the owners. The group revises the plan with information from bankers and investors once the project situation becomes more defined.

Although this difference is not as important for project development considerations, the feasibility study is only used prior to implementation. In contrast, businesses continue to use and revise their business plans after a project has been implemented.

The feasibility study refines the group's initial ideas, while the business plan uses information from the study to further prepare the project to evolve into an operating business.

Appendix C. Sample Feasibility Consultant Selection Criteria[4]

	Points		Awarded
Previous experience creating feasibility studies		(0-20)	
Knowledge of the industry to be studied		(0-15)	
Qualifications of principal researchers or team		(0-10)	
Understanding of the cooperative structure		(0-10)	
Proposed interaction with designated members		(0-15)	
Verbal presentation/communication skills		(0-10)	
Reasonableness of cost		(0-15)	
Miscellaneous intangible		(0-5)	
Total Score		100	

Appendix D. USDA Rural Development Summary Guide for Feasibility Studies Included in Applications for Business & Industry Loan Guarantees (Instruction 4279-B)

A feasibility study by a recognized independent consultant may be required by the Agency for start-up businesses or existing businesses when the project will significantly affect the borrower's financial operations.

[4] Adapted from USDA's Cooperative Service Report 54, "Creating 'Co-op Fever': A Rural Developer's Guide to Forming Cooperatives."

An acceptable feasibility study should include, but not be limited to:

a) Economic feasibility. Information related to the project site; availability of trained or trainable labor; utilities; rail, air, and road service to the site; and the overall economic impact of the project.
b) Market feasibility. Information on the sales organization and management, nature and extent of market and market area, marketing plans for sale of projected output, extent of competition, and commitments from customers or brokers.
c) Technical feasibility. Technical feasibility reports shall be prepared by individuals who have previous experience in the design and analysis of similar facilities or processes proposed in the application. The technical feasibility reports shall address the suitability of the selected site for the intended use including an environmental impact analysis. The report shall be based upon verifiable data and contain sufficient information and analysis so that a determination may be made on the technical feasibility of achieving the levels of income or production that are projected in the financial statements. The report shall also identify any constraints or limitations in these financial projections and any other facility or design-related factors which might affect the success of the enterprise. The report shall also identify and estimate project operating and development costs and specify the level of accuracy of these estimates and the assumptions on which these estimates have been based. For the purpose of the technical feasibility reports, the project engineer or architect may be considered an independent party provided neither the principals of the firm nor any individual of the firm who participates in the technical feasibility report has a financial interest in the project, and provided further that no other individual or firm with the expertise necessary to make such a determination is reasonably available to perform the function.
d) Financial feasibility. An opinion on the reliability of the financial projections and the ability of the business to achieve the projected income and cash flow. An assessment of the cost accounting system, the availability of short-term credit for seasonal business, and the adequacy of raw materials and supplies.
e) Management feasibility. Evidence that continuity and adequacy of management has been evaluated and documented as being satisfactory.

Appendix E. Sample Pro Forma Cash Flow

Pro forma cash flow statement, FY 20XX*

Item	Mth 1	Mth 2	Mth 3	Mth 4	Mth 5	Mth 6	Mth 7	Mth 8	Mth 9	Mth 10	Mth 11	Mth 12	Totals
CASH RECEIPTS													
Cash sales**													
Credit collections													
Commission fees													
Interest income													
Loans/equity													
TOTAL RECEIPTS													
CASH PAID OUT													
Purchases**													
Salaries													
Employee wages													
Payroll expense													
Bad debts													
Outside services													
Supplies													
Repairs & maintenance													
Advertising/promotion													
Car/travel													
Accounting & legal													
Rent													

Item	Mth 1	Mth 2	Mth 3	Mth 4	Mth 5	Mth 6	Mth 7	Mth 8	Mth 9	Mth 10	Mth 11	Mth 12	Totals
Telephone													
Utilities													
Insurance													
Property taxes													
Other taxes													
Interest on loans***													
Depreciation													
Miscellaneous													
Subtotal													
Principal payment***													
Capital purchases													
Income taxes													
Other withdrawl													
TOTAL CASH PAID													
CHANGE IN CASH													
Beginning balance													
Ending balance													

*May have multiple statements for different years.

**May have more than one cash sale and purchases line (more commodities/products).

***May have more loan and interest payment lines if more loans are obtained.

APPENDIX F. SAMPLE PRO FORMA INCOME STATEMENTS

Pro forma income statements, FY 20XX - FY 20XX*

Item	FY 20XX	FY 20XX	FY 20XX	FY 20XX	FY 20XX
INCOME	$	$	$	$	$
Cash sales					
Commission fees					
Total sales					
Cost of goods sold					
GROSS MARGIN					
EXPENSES Salaries					
Employee wages					
Payroll expense					
Bad debts					
Payroll expense					
Outside services					
Supplies					
Repairs & maintenance					
Advertising/promotion					
Car/travel					
Accounting & legal					
Rent					
Telephone					
Utilities					
Insurance					
Property taxes					
Other taxes					
Depreciation					
Miscellaneous					
TOTAL OPERATING EXPENSES					
Operating income					
Interest expense					
NET MARGIN					
Unallocated earnings					
Allocated earnings					

* This example shows five-year projections but many projects focus on just three years. Operating statement line items will vary in description and inclusion depending on project.

APPENDIX G. SAMPLE PRO FORMA BALANCE SHEETS

Pro forma balance sheets, FY 20XX - FY 20XX*

Item	FY 20XX	FY 20XX	FY 20XX	FY 20XX	FY 20XX
ASSETS	$	$	$	$	$
Current assets					
Cash					
Accounts receivable					
Inventory					
Prepaids (e.g., insurance)					
Other					
Total current assets					
Fixed assets					
Machinery & equipment					
Buildings					
Land					
Less: accumulated depreciation					
Total fixed assets					
TOTAL ASSETS					
LIABILITIES AND MEMBER EQUITY					
Current liabilities					
Accounts payable					
Taxes payable					
Patronage refunds payable					
Line of credit					
Interest payable					
Total current liabilities					
Long term liabilities					
Machinery and equipment note					
Real estate and building					
Total long term liabilities					
Total liabilities					
Member Equity					
Common stock					
Preferred stock					
Allocated earnings					
nallocated earnings					
Per unit capital retains					
Total member equity					
TOTAL LIABILITIES & MEMBER EQUITY					

* This example shows five-year projections but many projects focus on just three years. Balance sheet line items will vary in descriptions and inclusion depending on project.

APPENDIX H. SAMPLE PRO FORMA RATIO ANALYSIS

Pro forma financial ratio analysis, FY 20XX - FY 20XX*

Item	FY 20XX	FY 20XX	FY 20XX	FY 20XX	FY 20XX
Current ratio (current assets/current liabilities)					
Debt ratios (total debt/total assets)					
(total debt/member equity)					
Average collection period					
(receivables/sales per day)					
Total assets turnover (sales/total assets)					
Profitability ratios					
Return on equity (net margins/total equity)					
Return on Invesment (net margins/Investment)					
Return on sales (net margins/sales)					

* This example shows five-year projections but many projects focus on just three years. Ratio analysis items will vary in descriptions and inclusion depending on project.

In: Cooperatives
Editors: J. Rogers and V. K. Grant
ISBN: 978-1-62081-251-8

Chapter 2

EDUCATION INITIATIVES OF FARMER COOPERATIVES*

James J. Wadsworth

ABSTRACT

This report describes farmer cooperatives' education initiatives by providing information on whom they are educating and how, topics covered, their sources for educational materials and programs, and what educational topics they feel directors most need. The degree of cooperative education provided is reported for respondents overall and by cooperative type, size, and location.

PREFACE

This report examines how farmer cooperatives promote co-op education for members, directors, management, employees, and the general public. It provides information about whether cooperatives provide education, the audiences they reach, the delivery methods they use, the topics covered, where they get their educational resources, whether they feel there are enough

* This is an edited, reformatted and augmented version of United States Department of Agriculture, Rural Business–and Cooperative Programs, Publication Research Report 223, dated September 2011.

resources available, and what topics they feel are most important for educating directors.

The report provides the findings of the analysis, discusses implications of the results, and provides some recommendations for cooperatives and cooperative educators given the findings and implications.

Data for this study were collected from the U.S. Department of Agriculture (USDA) annual statistical survey (for fiscal year 2009) of U.S. farmer cooperatives. The cooperatives were asked to answer an additional seven questions about their cooperative education efforts. These data were sorted by whether they provide education and by cooperative type, size, and location. Of the population of 2,389 cooperatives surveyed, 31 percent, or 751, responded to the educational initiative section.

The author gratefully acknowledges college intern Paris Carter for her assistance with this study.

HIGHLIGHTS

This study provides a national perspective of cooperative education initiatives by farmer cooperatives. The analysis examines whom cooperatives educate, what resources they use, what topics they focus on, where they source their programs and materials, whether they feel there are enough resources available for education, and what educational topics they feel directors most need to fulfill their duties.

A USDA survey asked 2,389 farmer cooperatives to provide information on their educational initiatives with 34 percent (751) responding.

Some major findings include:

- Cooperative education is provided by 85 percent of respondents.
- More than 80 percent of grains and oilseeds, farm supply, service, and cotton and cotton gin cooperatives provide co-op education. More than 70 percent of dairy, fruits, vegetables, and nuts cooperatives provide it.
- Larger cooperatives provide education more than the smallest cooperatives.
- Directors, employees, and management are the primary audiences educated.
- Larger cooperatives educate directors more than other audiences. Smaller cooperatives vary in whom they educate most.

- Cooperatives educate by using staff more than any other method. External education through co-op councils or organizations that offer workshops is the second most used method.
- Cooperatives indicated that the general public, members, and employees need education the most; directors were next and management last.
- About 76 percent of cooperatives feel that there are enough educational resources available; 17 percent did not feel there are enough, and 8 percent didn't know.
- Cooperatives most often use cooperatives councils, USDA, and "other" cooperatives to acquire educational materials and programs.
- Cooperatives feel that directors need education in cooperative finance the most, followed by tax and legal issues, leadership, governance, board meeting functions, and lastly cooperative basics.

Some questions arise from the results of this study:

- What can be done to motivate cooperatives to educate their members more, and what educational initiatives or materials do cooperatives need to educate members?
- Are the educational activities cooperatives are undertaking adequate enough to educate their target audiences?
- Are the various educational delivery methods being used sufficient to reach all audiences needing cooperative education?
- Are cooperatives using internal staff to develop educational materials and development, or are these staff simply finding educational resources for the cooperative to use?
- Why aren't some types of cooperatives using cooperative council educational programs more?
- Why aren't some audiences receiving as much education as others (even when cooperatives feel that they need it)?
- Are there enough educational programs and materials available to meet all cooperative education needs? Are there enough resources, but some cooperatives just don't know about them, or use them?
- Are cooperatives receiving most of the resources that they truly need? Are there other resources that their major sources need to develop?
- Are cooperatives trying to educate the public on a regular basis and turning to certain sources to help with it? Or do they consider that

education to be an endeavor left to other cooperative educators to carry out?
- Are director education efforts sufficient? Are cooperatives regularly evaluating their director educational needs?

Some recommendations include:

- More cooperatives should become involved in cooperative education.
- Cooperatives need to ensure that the education conducted is extensive, and germane to the pertinent topics that their directors (and other audiences) need.
- Cooperatives need to educate members by providing more materials/programs for them and/or holding special member events.
- Cooperatives must actively search for educational materials and programs.
- Cooperatives need to focus their educational activities to closely match their perceptions of where education is most needed.
- Cooperatives need to freely reach out to educators when they feel gaps exist in materials and programs, and educators need to work to fill those gaps.
- Educators need to effectively promote their materials and programs to ensure that cooperatives are aware of and can gain access to them.
- Researchers should conduct further study of cooperative educational initiatives to survey the actual content of cooperatives' educational undertakings to allow for development of more targeted and effective education materials and programs.

INTRODUCTION

Education is the lifeblood of cooperatives worldwide. The fifth International Cooperative Principle states:

Education, Training, and Information—Cooperatives provide education and training for their members, elected representatives, managers and employees so they can contribute effectively to the development of their cooperative. They inform the general public—particularly young people and opinion leaders—about the nature and benefits of cooperation.[1]

The cooperative business model is unique, and those involved in it—members, directors, employees, and management—must understand all its nuances. The general public also needs to understand how cooperatives differ from other business types and how cooperatives' guiding principles come into play in their lives and communities.

Cooperative educators have long focused on certain audiences for cooperative education work (e.g., workshops, programs, publications, materials, etc.). Directors and management have been a key audience, as well as those interested in learning how to develop a new cooperative. In 2009, a good number of cooperative educators indicated that people interested in forming cooperatives, directors, and management were the three top audiences to educate. [2]These educators also pointed to finance as the most crucial education topic. The goal of this study was to gain a better understanding of what farmer cooperatives are undertaking for their educational initiatives. This report consists of four major sections: information about the survey instrument used, discussion of the statistical findings, implications and questions pertaining to the results, and conclusions and recommendations.

SURVEY DESIGN AND RESPONSE

Data for the study were collected using USDA's annual survey of farmer, rancher, and fishery cooperatives. In addition to the usual annual financial questions, questions pertaining to education were asked in the 2010 survey (for fiscal 2009). The additional questions asked whether the cooperatives provide education to various audiences (i.e., members, directors, employees, management, and the general public), and, if so, by what means. The cooperatives were also asked: to rate the audiences relative to their need of cooperative education; whether they feel there are enough education resources available; where they get their educational programs or materials; and to rate the most crucial education topics for directors (see Appendix Note 1). Assessing all the answers together for each respondent provided a view of their education initiatives. Only those surveys well filled out were included in the analysis for that reason. Of the 2,389 U.S. farmer cooperatives sent surveys, there were 751 (31 percent) usable responses to the education survey. The sample appears to be a fairly well represented subset of the total population.[3] The annual survey of farmer cooperatives collects other data (i.e., financial, operational, structural, etc.). Those data corresponding to the cooperative education survey respondents were combined with the education data. This allowed for cross tabulation of education results by cooperative type, size, location, and financial position.

Results-Cooperative Education Initiatives and Aspects

Farmer cooperatives provide education on cooperative principles, finance, governance, tax and legal issues, and a host of other cooperative-related topics. The amount of education conducted and for whom, however, is not well known for all cooperatives. In this study, cooperative education initiatives were analyzed by overall sample and by size, type, and location. Cooperatives and cooperative educators may find it useful to further assess specific areas of information in which they are interested from the tables contained herein.

Respondents Provision of Education

A majority (85 percent) of the respondent cooperatives to the survey said that they provide education to members, directors, employees, management, or the general public (Table 1). Table 2 shows the provision of education by respondent type (see Appendix note 3 for more on cooperative type). About 92 percent of grain and oilseed respondents provide education, followed by farm supply and service cooperatives at 85 percent. Cotton and cotton gin cooperatives are at 81 percent; dairy at 79 percent; fruits, vegetables, and nuts at 74 percent; livestock, poultry, and wool cooperatives at 63 percent; and "other" marketing cooperatives at 65 percent. Larger cooperatives appear to educate more than smaller cooperatives (Table 3). More than 90 percent of cooperative respondents in the three largest size categories provide education, while only 59 percent of the smallest cooperative respondents do. Table 4 provides data categorized by location. Six regions (made up of various states) are depicted— Pacific, Intermountain, Great Plains, Southeast, North Central, and Northeast. Data from some select states are also shown for greater information, but the state analyses were limited to only those states that had at least 15 responses to most survey questions. The cooperative respondents in the Pacific, Great Plains, and North Central regions provide the most education, with 88 percent, 87 percent, and 86 percent, respectively, providing education. The Northeast region respondents provide the least amount of education, with a 62-percent provision rate. Seventy-nine percent of the respondents from both the Intermountain and Southeast regions provide education. Respondents from Nebraska had the highest provision rate at 97 percent, but there are also six other states with education provision rates of more than 90 percent—Washington, South Dakota, Tennessee, Iowa, Ohio, and Wisconsin. The five states of California, Kansas, Oklahoma, Texas, and Minnesota had provision rates of more than 80 percent. Of the states included in the analysis, those with the lowest rates include Missouri, Michigan, Pennsylvania, and Montana.

Audiences Receiving Education

For those respondents that provide education (Table 5), most indicated that they provide it mostly to directors (82 percent), followed by employees (70 percent), management (66 percent), members (35 percent), and the general public (15 percent). Table 6 shows the audiences that cooperative respondents educate, by type of cooperative (see Appendix Note 3 for a definition of the cooperative types discussed in this report). Farm supply and grain and oilseed cooperatives provide education mostly to directors and employees. Dairy cooperatives educate directors and members more than other audiences. Fruits, vegetables, and nuts cooperatives educate directors and management the most. Cotton and other marketing cooperatives educate directors the most, then employees. Livestock cooperatives educate members the most, while service cooperatives educate directors and management to the same degree. Overall, directors are the predominate audience educated, with the only exception being in livestock, poultry, and wool cooperatives. More larger cooperatives than smaller provide education to directors, with 97 percent of cooperatives with assets of $50 million or more providing education to this audience (Table 7). Among the smallest cooperative respondents, members were the audience receiving the most education, followed by directors. The general public is the audience receiving the least education by cooperative respondents of all sizes, while directors, employees, management, and members were more favored for education. Table 8 shows audiences that cooperatives educate by region and select states. The cooperatives in the Pacific, Intermountain, Great Plains, and North Central regions all put most of their efforts in education for directors. The Pacific and Intermountain regions had management as the second audience they educate, while the Great Plains and North Central regions had employees second. The Southeast region educates employees most, followed by directors, and the Northeast cooperatives educate members first with directors second. Respondent cooperatives in 11of the 17 states educate directors the most. Texas, Pennsylvania, Tennessee, and Wisconsin educate directors and employees in equal amounts. South Dakota respondents educate employees the most, Missouri educates management, and Pennsylvania educates members.

Delivery of Education

For the education that respondents do provide, most use several means of delivery. Table 9 shows the breakdown of delivery methods used. Sixty-four percent of the respondents indicated that they provide education internally with their staff. About 60 percent said they get education to their audiences by attending workshops provided by a cooperative council or another cooperative organization. Other means of delivery that have significant use includes internal publications and materials, attendance at national conferences, and internal use of a consultant. It is apparent that many cooperatives use a variety of methods for delivering education to their audiences and do not rely on just one method. Delivery methods vary by type of cooperative, but use of cooperative staff, cooperative councils, and publications and materials were predominately the methods used (Table 10).

Farm supply, dairy, livestock (wool and poultry), other marketing, and service cooperatives all used internal staff the most. Grain and oilseed, fruits (vegetables and nuts), and cotton cooperatives used cooperative council programs the most.

Table 11 shows delivery methods by cooperative size. Similar to type, use of cooperative staff, cooperative councils, and publications and materials are the major delivery methods across cooperative sizes, except for those in the \$20 to \$49.9 million group, which use cooperative councils most. Use of cooperative staff is the most used delivery method by cooperatives of all sizes. Cooperative respondents larger than \$2.5 million in assets reported that they used cooperative staff and cooperative councils the most for delivering education.

The smallest cooperatives (under \$2.5 million) used cooperative staff the most, and then publications and materials for delivery. For the most part, cooperative councils are a major education delivery vehicle for cooperatives that reside in council states or regions (Table 12).

For instance, 85 percent of the cooperatives in Kansas, 83 percent of those in Iowa, and 79 percent of those in Nebraska indicated that they get their cooperative education delivered through cooperative councils, in addition to other major methods such as use of cooperative staff and publications and materials.

Educational Need of Key Audiences

Cooperatives were polled on which audiences they think need education. Table 13 shows that cooperatives perceive that the general public most needs cooperative education, followed respectively by members, employees, directors, and management. Overall, the respondents feel that all the audiences are in need of cooperative education to some degree. Farm supply cooperatives indicated that the general public needs education the most (Table 14). Dairy and livestock, poultry, and wool cooperatives said that members need education the most. Grains and oilseeds cooperatives had a tie between general public and members for needing education the most. The highest rating for all audiences by type was a 4.1 (on a scale of 1 to 5 with 1 needing least and 5 needing most) by other marketing cooperatives for directors—thus, those cooperatives feel that directors need education the most. The next highest rating was 3.7 for the general public by farm supply cooperatives. The lowest rating was 1.5 by other marketing cooperatives for the general public. Overall, management was perceived as needing education the least by most cooperatives. Cotton and cotton gin cooperatives were the exception as they indicated that management needed education most. Table 15 shows that the general public was the audience perceived to need education on cooperatives the most by all but the second smallest respondent size category. Members were also scored relatively high in the perceived need of education, by all the size categories. Respondent cooperatives in most of the regions and states included in Table 16 thought that the general public needed education about cooperatives the most. Cooperatives from 16 regions or states rated the general public the highest in need (although two had ties between general public and members). Respondents in seven regions or states rated members highest, while three rated directors highest (although three had ties between members and other audiences), and respondents two states had employees rated highest, though tied with members. However, the ratings for most of these three audiences-members, directors, and employees-did not differ by a great deal in most cases.

Educational Resources

More than 75 percent of the respondents indicated that there are enough education resources, such as materials, programs, workshops, conferences, etc., available to them for educational needs (Table 17). About 17 percent said there are not, and almost 8 percent did not provide an opinion. The perception of the availability of cooperative educational resources differs somewhat by cooperative type. About 82 percent of fruits, vegetables, and nuts cooperatives, as well as the same proportion of grain and oilseed cooperatives, think there are enough resources (Table 18). Other types where the majority feels there are enough resources include farm supply, cotton, service, and dairy. However, less than 50 percent of livestock, poultry, wool and "other" marketing cooperative respondents indicated that there are enough resources available. Table 19 presents perception of availability of educational resources by size. More than 70 percent of all the size categories, except the smallest group (assets less than $1 million), indicated they thought there were enough resources. Just 52 percent of the smallest size felt there are enough. Also, 26 percent of the largest size group ($50 million or more in assets) indicated that there are not enough education resources available. The Great Plains region was the only region where more than 80 percent of the respondents said that there are enough education resources available (Table 20). However, eight states had over 80 percent saying there are enough: California, Kansas, South Dakota, Tennessee, Iowa, Michigan, Ohio (over 90 percent), and Wisconsin. The regions that had the highest proportions (more than 20 percent) saying that there aren't enough resources are the Intermountain, the Southeast, and the Northeast, while the highest proportions among states are Washington, Montana, Oklahoma, Minnesota, Missouri, and Pennsylvania.

Sources of Educational Resources Used

Table 21 shows that respondents who conduct education acquire their resources largely from the state or regional cooperative councils (58 percent), from USDA (43 percent), and from other cooperatives (38 percent). A variety of other sources are also used, such as the Internet (25 percent), universities, and cooperative centers (both 15 percent). Cooperative development centers are not yet a significant source, with only 5 percent of respondents noting that they attain resources from them. Other sources (e.g., National Society of Accountants for Cooperatives, attorneys, auditors, consultants, National Farmers Union, Farm Bureau, trade organizations, etc.) were attained by 12 percent of cooperatives. By type, co-op councils are most widely used by grain and oilseed (67 percent), farm supply (58 percent), and cotton and cotton gin (70 percent) cooperatives (Table 22). USDA is most widely used by dairy (58 percent); fruits, vegetables and nuts (53 percent); and livestock, poultry, and wool (83 percent) cooperatives. Besides cooperative councils and USDA, other fairly significant sources for the various types are other cooperatives, universities, cooperative centers, online resources, and other sources. Table 23 shows that smaller cooperatives (up to $5 million in assets) rely more heavily on USDA to attain educational resources, and larger cooperatives (greater than $5 million in assets) rely more on the state or regional cooperative councils. Other cooperatives are fairly well used also. Cooperatives of all sizes also use universities, online information/courses, cooperative centers, etc., as sources. Co-op councils and USDA were the primary resources for educational materials in 13 states and 5 regions, respectively (Table 24). USDA was the second most used in nine regions.

Importance of Director Educational Topics

Cooperatives were asked to rate the importance of director education on six overall topics: cooperative basics, finance, governance, leadership, board meeting functions, and tax and legal issues. The highest need rating was for cooperative finance, followed by (in highest order of need) tax and legal issues, leadership, governance, board meeting functions, and cooperative basics (Table 25). The highest rating was 3.4 for finance (on the scale of 1 to 5 with 1 being least needed and 5 being most needed) and the lowest was 2.8 for basics. By type, director training in finance is rated as being most needed by farm supply, grains and oilseed, dairy, service, and other marketing cooperatives (Table 26). Fruits, vegetables, and nuts and livestock, poultry, and wool cooperatives rate basics as most needed education for directors. Leadership, board meetings, and tax and legal topics are equally rated highest by cotton and cotton gin cooperatives. Table 27 shows that finance rated high in relative terms of director education need by all classes of cooperative size, from small to large. Governance, leadership, and taxes and legal are rated relatively high in need also. Basic cooperative education is rated relatively high by the two smaller cooperative groups. By area, cooperative respondents in the Great Plains, Intermountain, Southeast, North Central, and Northeast regions and in the states of South Dakota, Iowa, Michigan, and Minnesota rated finance as highly needed for director education (Table 28). There were also five other regions and states—Great Plains, Northeast, North Dakota, Texas, and Wyoming—that rated finance as highest although it was tied with another topic. Board meeting functions is the topic rated highest by cooperatives in the Pacific and California (although California also had the same rating for cooperative basics and governance). Leadership as a topic is rated high by cooperatives in Washington, Nebraska, Oklahoma, Tennessee, and Wisconsin. The tax and legal issue topic is highly rated among cooperatives in the Great Plains, Nebraska, North Dakota, and Texas. Basic cooperative education is rated highest in need by cooperatives in California, Washington, Missouri, Ohio, and Pennsylvania.

Select Findings by Top Educational Sources Used

Table 29 provides select educational findings for those respondents that noted that they source their educational programs and materials through cooperative councils, USDA, and other cooperatives (the top three sources of respondents). Some of the respondents are included in all three categories. Of the 389 cooperatives that source materials/programs from state or regional councils, 45 percent (176 cooperatives) also use USDA as a source and 35 percent (138 cooperatives) use other cooperatives. Of the 287 cooperatives that use USDA, 37 percent (107 cooperatives) also use other cooperatives. Two hundred fifty-four cooperatives use other cooperatives as a source. Cooperatives using all three of these sources educate directors more than other audiences. Ninety-one percent of the cooperatives that use cooperative councils educate directors, while 84 percent and 88 percent that use USDA and other cooperatives, respectively, educate directors. More than 70 percent of cooperatives using the three sources educate employees and management. Cooperatives using all three sources rated the general public as the audience that most needs cooperative education. Members and employees were rated second and third, respectively, by these cooperatives for need of education. More than 80 percent of cooperatives in all three source groups feel that there are enough education resources available, higher than the 75 percent of the entire population of respondents.

Cooperatives using cooperative councils as an education source rated finance as the education topic most needed by directors (rating of 3.4), followed by leadership, and tax and legal education (both rated 3.2). Governance education was next (rating of 3.1). Cooperatives using USDA rated finance and tax and legal topics highest (ratings of 3.3), followed by leadership (rating of 3.2).

Governance was next (rating of 3.1). Cooperatives using other cooperatives rated finance highest (rating of 3.4), followed by tax and legal (rating of 3.2). Governance and leadership topics were next (ratings of 3.1).

Financial Position

Table 30 presents aggregate financial data of the 751 respondent cooperatives that provided information on their educational initiatives. The 636 cooperatives that provide education (educators) had average total sales of $86.7 million, while the 115 cooperatives that do not provide education (non-educators) had total sales of $31.5 million. The educators' average net income was $2.2 million, and the non-educators' average net income was $0.96 million. The average educator cooperative had a 9.1—percent return on assets, and non-educator cooperatives had 6.4 percent. Educators were more highly leveraged than non-educators. The educator cooperatives had an average of 76 full-time employees and 950 members, while the non-educator cooperatives had 29 full-time employees and 564 members.

Study Implications

This section provides an analysis of some of the implications of the study results. Some questions that educators would likely ask given the derived implications are presented also.

A Majority of Cooperatives Educate

The 85-percent majority of the respondent cooperatives that provide education are a fairly significant proportion, but 100 percent should be the goal. That 90 percent of grain and oilseed cooperatives are providing education, and that 85 percent of farm supply and service cooperatives do, is commendable. But 25 percent or more of some types of cooperatives are not providing any education, which is not as positive. Larger cooperatives provide more education than smaller cooperatives likely because they simply have more resources. Larger cooperatives may also be more aware of cooperative education materials, publications, and programs and have a closer tie to cooperative councils, universities, and other resources than smaller cooperatives do, making it easier for them to provide education. While none of the six regions had 90 percent or more of the associated cooperatives providing education, seven of the states did. It is interesting that all of those states have active cooperative councils providing education, likely not a coincidence. All but one of the five states with more than 80 percent of the cooperatives involved in educating also have councils active in education. California is the exception (while California does have an active cooperative council, it does not conduct educational activities).

Director Education Important

Directors are the audience being most educated by cooperatives. Given the complex business environment in which cooperatives operate, keeping directors up to speed on cooperative and business issues is imperative, and cooperatives seem to understand this. The largest cooperatives are educating directors the most, likely because of the complexity in their operations that their directors need to understand. Of all audiences, members are the least educated by cooperatives, receiving even less education from respondents than the general public does. But there is an exception—livestock, poultry, and wool cooperatives do educate members more than other audiences. This could be because those cooperatives have strong competition with non-cooperatives and, thus, strive to keep members well aware of cooperative benefits. Still, cooperative leaders often voice that loyalty and commitment from their members is a major issue they face, so it would benefit them to improve that by educating members more than they currently do. A few questions arising from these implications include: what can be done to motivate cooperatives to educate their members more, and what educational initiatives do cooperatives need to use to educate members? Are the educational activities cooperatives are undertaking adequate enough to fully educate their target audiences?

Education Delivery Methods Raise Several Questions

The finding that cooperatives provide most of their education "internally with cooperative staff" begs the question: what type of education are these staffs providing? Are they developing their own materials and programs, or simply finding educational resources for the cooperative to use? Cooperative council educational programs are fairly well used, with 60 percent of cooperatives using them for education delivery. Grain and oilseed and cotton and cotton gin cooperatives use them more than other types of cooperatives. In light of this, why aren't livestock, poultry, wool, and dairy cooperatives using council programs more than they are? Part of the answer is likely council membership status or location proximity. Some of those cooperatives not using councils may not be members, or location logistics do not warrant easy participation in council programs. Overall, cooperatives clearly use multiple education methods. A question arises given this finding: are the various delivery methods being used often enough and being targeted to reach all the key audiences needing cooperative education?

Perception of Audience Needs Doesn't Match Education Conducted

All audiences are perceived to need education to some degree by cooperatives. An interesting finding is that the general public was deemed as needing education the most. However, only a small proportion of cooperatives are providing education to the general public relative to other audiences. Members were also perceived as needing more education, but they too, are not receiving it from most cooperatives. Thus, two audiences felt to be needing education are not receiving much of it from cooperatives—the question is why not? It may be that cooperatives think that the general public doesn't understand their business very well, or maybe boards don't feel it's that important, or financially worthwhile, to educate them. Cooperatives may need to focus their educational efforts to more closely match their perceptions of need when they feel it will benefit them. The finding that farm supply cooperatives feel that the general public needs education most (even more than all the other cooperative types) may imply that farm supply cooperatives do not feel that the general public understands that their supply business is different (i.e., owned by farmers) than other private farm supply businesses, and, thus, they think that educating them on that is very important.

Resources Not Perceived by Some as Being Widely Available

While 85 percent of the cooperatives provide education, 75 percent feel that there are enough educational resources available. This finding brings up some interesting questions. Are there not enough resources available to meet all cooperative education needs in the United States? Are there enough resources, but some cooperatives just don't know about them, or use them? Are there resources, but not entirely of the right kind to meet all needs? These questions signal that there are some issues with cooperative educational resources that need to be addressed—that either resources are not plentiful enough to fully meet needs, or the promotion strategies of the many resources that are available are not far reaching enough.

Multiple Sources Used for Educational Resources

Cooperatives that conduct education use a variety of resources for their material and programs. Cooperative councils, USDA, and other cooperatives were most used by respondents. It is highly likely that through the use of cooperative councils, cooperatives are acquiring several types of resources—educational workshops and conferences, as well as some informative type literature in hard copy or electronic format. For those using USDA, it can be readily assumed that the cooperatives are gaining access to its reports, publications, magazine articles, and perhaps some of its electronic materials containing education programs or information (e.g., CDs, VHS, and DVDs). For those acquiring resources from other cooperatives, universities, online, etc., it's a bit more difficult to ascertain what it is they are using or getting. Some cross-over acquisition is going on. For example, a cooperative council may use USDA educational publications for its educational workshops or handouts. Or it might use a professor for a speaker or instructor, or university studies and materials. Together with the finding that many cooperatives feel there are not enough resources for education, questions arise as to what cooperatives are actually getting from their sources. Are they receiving most of the resources that they truly need?

Are there other resources that their major sources need to provide? Hopefully, cooperatives voice their needs to those they use as sources. Feedback and communication are extremely important in the realm of providing cooperative educational programs and materials. For those cooperatives using these three main sources, directors, management, and employees are educated the most—however, these cooperatives indicated that the general public is the audience that needs the most educating. So, the question is, are cooperatives trying to educate the public on a regular basis and turning to certain sources to help with it? Or do they consider that education to be an endeavor left to other cooperative educators to carry out?

A higher majority of the cooperatives who use the three major sources feel there are enough educational resources than the overall responders do, but nearly 20 percent of this majority still feel that there are not enough. Thus, it is incumbent upon educators and the cooperatives themselves, in some instances, to find where the gaps are and fill them.

All Topics Are Important for Director Education, but Finance Needs Stand out

Cooperative respondents feel finance is the educational topic most needed by directors. However, some types of cooperatives rated other topics higher. Overall, the findings reflect that all the topics are felt to be needed for director education.

These findings were similar among the cooperatives that sourced cooperative councils, USDA, and other cooperatives as well. As farmer cooperatives strive to operate effectively in a tough economic climate, well-educated directors are an imperative to sound governance. Questions arise from these topic findings: cooperatives indicate early on in the survey that they are educating directors, but are they educating them enough on the various topics they feel they need? Are cooperatives regularly evaluating their director educational needs so that they can ensure directors are getting the needed education?

Financial Meaning

The financial data shown for respondents, broken down into whether they conduct education, does not provide any significant implications. However, the data reinforces the finding that larger cooperatives provide more education, with average total assets and total sales of educators being $20 million and $55 million more, respectively, than non-educators.

Performance is only slightly higher for educators relative to non-educators, and both categories performed well financially, as did the entire sample population.

CONCLUSION AND RECOMMENDATIONS

Cooperatives providing education on their business model, its unique characteristics, and the environment in which they operate, is extremely important. Many audiences need to be reached so that a full understanding of cooperatives is gained.

Members need to know more about their business; directors need to understand complex issues and cooperative principles and practices to properly govern; managers need knowledge and tools to effectively manage the operations; employees need to understand cooperatives in order to properly serve member-owners; and the general public should understand how cooperatives differ from other business forms and how they benefit their communities so they can fully appreciate and support them. This study provides information on educational initiatives by farmer cooperatives—whom they educate, how they do it, what resources they use, which audiences they feel need educating the most, and the topics on which they think directors need to be educated.

However, due to space restrictions, the survey instrument used was limited in the number of questions it could include. More questions could have delved deeper into cooperatives' specific education activities to enhance the results of this study.

Nevertheless, this study provides some interesting information that should be useful to cooperatives and educators as they develop cooperative educational activities. Many cooperatives provide cooperative education and take advantage of available resources. However, some cooperatives do not educate at all. Some recommendations for farmer cooperatives and educators given the results of this study include:

- Cooperatives that aren't educating any audiences should strive to become educators.
- Cooperatives need to make sure that the education conducted is extensive and germane to the pertinent topics that their directors (and other audiences) need.
- Cooperatives need to educate members by providing more materials/programs for them and/or holding special member events.
- Cooperatives must actively search for educational materials and programs.
- Cooperatives need to focus their educational activities to closely match their perceptions of where education is needed.
- Cooperatives need to freely reach out to educators when they feel gaps exist in materials and programs, and educators need to work to fill those gaps.
- Educators need to effectively promote their materials and programs to ensure that cooperatives are aware of and can gain access to them.
- Researchers should conduct further study of cooperative educational initiatives to survey the actual content of cooperatives' educational undertakings to allow for development of more targeted and effective education materials and programs.

Table 1. Cooperative respondents' indication of providing cooperative education

	Number	Perce
Yes	636	84.7
No	115	15.3
Total	751	100

751 respondents represent 31 percent of the 2,389 farmer cooperatives in the United States.

Table 2. Cooperative respondents' indication of providing cooperative education, by type

Cooperative type	Yes		No
		percent	
Farm supply (n = 345, 45.9%)	84.6		15.4
Grain and oilseed (n = 225, 30%)	91.6		8.4
Dairy (n = 47, 6.3%)	78.7		21.3
Fruits, vegetables, & nuts (n = 39, 5.2%)	74.4		25.6
Cotton & cotton gins (n = 36, 4.8%)	80.6		19.4
Livestock, poultry, & wool (n = 16, 2%)	62.5		37.5
Other marketing (n = 17, 2.3%)	64.7		35.3
Service (n = 26, 3.5%)	84.6		15.4
Total (n = 751, 100%)	84.7		15.3

See Appendix note 3 for a description of the cooperative type categories.

Table 3. Cooperative respondents' indication of providing cooperative education, by size

Cooperative asset size	Yes		No
		percent	
Less than $1 million (n = 75, 10%)	58.7		41.3
$1 million to $2.49 million (n = 74, 10%)	78.4		21.6
$2.5 million to $4.99 million (n = 125, 16.6%)	86.4		13.6
$5 million to $9.99 million (n = 139, 18.5%)	84.9		15.1
$10 million to $19.99 million (n = 142, 18.9%)	90.1		9.9
$20 million to $49.99 million (n = 113, 15%)	91.2		8.8
$50 million or more (n = 83, 11%)	92.8		7.2
Total (n = 751, 100%)	84.7		15.3

Table 4. Cooperative respondents' indication of providing cooperative education, by location

Cooperative location	Yes		No
		percent	
Pacific (n=68, 9.1%)	88.2		11.8
California (n = 33, 4.4%)	84.8		15.2
Washington (n = 21, 2.8%)	95.2		4.8
Intermountain (n=48, 6.4%)	79.2		20.8
Montana (n = 16, 2.1%)	75.0		25.0
Great Plains (n = 240, 32%)	87.1		12.9
Kansas (n = 53, 7%)	88.7		11.3
Nebraska (n = 34, 4.5%)	97.1		2.9
North Dakota (n = 53, 7%)	77.4		22.6
Oklahoma (n = 18, 2.4%)	88.9		11.1
South Dakota (n = 28, 3.7%)	93.1		6.9
Texas (n = 54, 7.2%)	85.2		14.8
Southeast (n = 58, 7.7%)	79.3		20.7
Tennessee (n = 24, 3.2%)	95.8		4.2
North Central (n = 303, 40.3%)	86.5		13.5
Iowa (n = 50, 6.6%)	92.0		8.0
Michigan (n = 18, 2.4%)	66.7		33.3
Minnesota (n = 84, 11.2%)	85.7		14.3
Missouri (n = 22, 2.9%)	59.1		40.9
Ohio (n = 19, 2.5%)	94.7		5.3
Wisconsin (n = 48, 6.4%)	93.8		6.2
Northeast (n = 34, 4.5%)	61.8		38.2
Pennsylvania (n=18, 2.4%)	72.2		27.8
Total (n = 751, 100%)	84.7		15.3

States included in each region: Pacific—CA, OR, WA, AK, and HI; Intermountain—AZ, NM, CO, UT, ID, WY, and MT; Great Plains—ND, SD, NE, KS, OK, and TX; Southeast—AR, LA, MS, AL, GA, FL, SC, NC, TN, KY, VA, and WV; North Central—MN, IA, MO, IL, WI, IN, OH, and MI; Northeast—MD, DC, DE, PA, NY, NJ, CT, RI, MA, VT, NH, and ME.

States included are those that had more than 15 responses (78.7% of respondents).

n = number that provided information (in this case, total number of respondents).

Table 5. Audiences of cooperative training or provision of educational materials concerning cooperative practices or issues

Members	34.9%
Directors	82.2%
Employees	70.0%
Management	66.3%
General public	14.8%

N = 636 (number that provided information).

Table 6. Audiences that cooperatives educate, by type

Cooperative type	Members	Directors	Employees	Management	General Public
			percent		
Farm supply (n = 292)	37.0	82.5	80.1	68.8	20.9
Grain and oilseed (n = 206)	27.7	87.9	73.3	68.0	10.7
Dairy (n = 37)	64.9	67.6	43.2	50.0	10.8
Fruits, veg. & nuts (n = 29)	48.3	86.2	31.0	75.9	6.9
Cotton & cotton gins (n = 29)	24.1	72.4	62.1	55.2	3.4
Livestock, poultry, & wool (n = 10)	70.0	60.0	10.0	30.0	10.0
Other marketing (n = 11)	9.1	90.9	36.3	81.8	
Service (n = 22)	18.2	63.6	54.5	63.6	13.6
Total (n = 636)	34.9	82.2	70.0	66.3	14.8

n = number that provided information.

Table 7. Audiences that cooperatives educate, by size

Cooperative asset size	Members	Directors	Employees	Management	General Public
			percent		
Less than $1 million (n = 44)	59.1	47.7	38.6	29.5	11.4
$1 million to $2.49 million (n = 58)	32.8	70.7	70.7	55.2	19.0
$2.5 million to $4.99 million (n = 108)	25.9	69.4	69.4	61.1	17.6
$5 million to $9.99 million (n = 118)	25.4	83.9	72.0	66.9	10.2
$10 million to $19.99 million (n = 128)	34.4	90.6	70.3	70.3	18.0
$20 million to $49.99 million (n = 103)	33.0	93.2	76.7	77.7	12.6
$50 million or more (n = 77)	53.2	97.4	75.3	80.5	14.3
Total (n = 636)	34.9	82.2	70.0	66.3	14.8

n = number that provided information.

Table 8. Audiences that cooperatives educate, by location

Cooperative location	Members	Directors	Employees	Management	General Public
			percent		
Pacific (n=60)	38.3	88.3	46.7	66.7	18.3
California (n = 28)	25.0	82.1	32.1	75.0	7.1
Washington (n = 20)	50.0	90.0	45.0	55.0	10.0
Intermountain (n=38)	44.7	81.6	65.8	68.4	13.2
Montana (n = 12)	50.0	75.0	58.3	50.0	8.3
Great Plains (n = 209)	31.6	78.9	73.2	66.5	16.3
Kansas (n = 47)	34.0	87.2	74.5	68.1	8.5
Nebraska (n = 33)	27.3	87.9	75.8	72.7	18.2
North Dakota (n = 41)	46.3	80.5	78.0	73.2	29.3
Oklahoma (n = 16)	25.0	81.3	62.5	68.8	6.3
South Dakota (n = 27)	40.7	66.7	74.1	48.1	29.6
Texas (n = 46)	17.4	67.4	67.4	63.0	8.7
Southeast (n = 46)	37.0	71.7	73.9	50.0	19.6
Tennessee (n = 23)	30.4	82.6	82.6	60.9	21.7
North Central (n = 262)	31.7	88.5	76.0	71.0	12.2
Iowa (n = 46)	32.6	91.3	76.1	76.1	23.9
Michigan (n = 12)	41.7	91.7	58.3	83.3	
Minnesota (n = 72)	26.4	90.3	72.2	72.2	12.5
Missouri (n = 13)	30.8	61.5	76.9	84.6	
Ohio (n = 18)	16.7	100.0	66.7	50.0	5.6
Wisconsin (n = 45)	42.2	91.1	91.1	80.0	11.1
Northeast (n = 21)	76.2	42.9	28.6	38.1	14.3
Pennsylvania (n=13)	84.6	38.5	38.5	23.1	15.4
Total (n = 636)	34.9	82.2	70.0	66.3	14.8

States included in each region: Pacific—CA, OR, WA, AK, and HI; Intermountain—AZ, NM, CO, UT, ID, WY, and MT; Great Plains—ND, SD, NE, KS, OK, and TX; Southeast—AR, LA, MS, AL, GA, FL, SC, NC, TN, KY, VA, and WV; North Central—MN, IA, MO, IL, WI, IN, OH, and MI; Northeast—MD, DC, DE, PA, NY, NJ, CT, RI, MA, VT, NH, and ME. States included are those that had more than 15 responses. n = number that provided information.

Table 9. Education delivery methods

Delivery	percent
Internally with cooperative staff	64.3
Internally using a consultant	30.2
Internally through use of educational publications and materials	38.5
Internally through an online course	7.6
Externally through a cooperative council or organization putting on a workshop	59.8
Externally through an online course	6.3
Joint educational meeting with another cooperative	15.4
Attendance at a state or national conference where education is provided	34.4
Other ways*	3.0

*Newsletters, mailings, local meetings, seminars, etc. N = 636.

Table 10. Education delivery methods, by type

	Internally with				Externally with			
Cooperative type	Co-op staff	Consultant	Pubs & materials	Online course	Co-op council	Online course	Joint meet	Attend conf
					percent			
Farm supply (n = 292)	71.2	29.8	43.8	8.2	55.8	4.8	18.5	37.3
Grain and oilseed (n = 206)	60.7	35.4	34.0	8.2	72.3	7.3	11.6	32.5
Dairy (n = 37)	67.6	16.2	40.5		29.7	5.4	5.4	32.4
Fruits, veg. & nuts (n = 29)	37.9	24.1	37.9	10.3	65.5	13.8	6.9	31.0
Cotton & cotton gins (n = 29)	48.3	24.1	20.7	3.4	72.4	6.9	17.2	34.5
Livestock, poultry, & wool (n = 10)	60.0	20.0	40.0	10.0	20.0	20.0	20.0	30.0
Other marketing (n = 11)	63.6	54.5	54.5	9.1	54.5		18.2	27.3
Service (n = 22)	59.1	18.2	22.7	4.5	40.9	4.5	31.8	27.3
Total (n = 636)	64.3	30.2	38.5	7.6	59.8	6.3	15.4	34.4

n = number that provided information.

Actual wording of delivery methods on survey:

Internally with cooperative staff, Internally using a consultant, Internally through use of educational publications and materials, Internally through an online course, Externally through a cooperative council or organization putting on a workshop, externally through an online course, Joint educational meeting with another cooperative, Attendance of state or national conference where education is provided.

Table 11. Education delivery methods, by size

		Internally with*			Externally with			
Cooperative asset size	Co-op staff	Consultant	Pubs & materials	Online course	Co-op council	Online course	Joint meet	Attend conf
	percent							
Farm supply (n = 292)	71.2	29.8	43.8	8.2	55.8	4.8	18.5	37.3
Less than $1 million (n = 44)	47.7	6.8	36.4	2.3	15.9	4.5	18.2	20.4
$1 million to $2.49 million (n = 58)	51.7	12.1	40.0	6.9	37.9	3.4	8.6	24.1
$2.5 million to $4.99 million (n = 108)	53.7	24.1	30.5	2.8	49.1	3.7	13.0	27.8
$5 million to $9.99 million (n = 118)	69.5	32.2	34.7	2.5	61.9	2.5	17.8	35.6
$10 million to $19.99 million (n = 128)	73.4	38.3	38.3	10.2	67.2	7.0	16.4	32.0
$20 million to $49.99 million (n = 103)	62.1	34.9	37.9	9.7	79.6	5.8	16.5	42.7
$50 million or more (n = 77)	77.9	42.9	57.1	18.2	74.0	18.2	15.6	50.6
Total (n = 636)	64.3	30.2	38.5	7.6	59.8	6.3	15.4	34.4

n = number that provided information.

*For actual wording of delivery methods on survey, see Table 10 footnote.

Table 12. Education delivery methods, by location

		Internally with			Externally with			
Cooperative type	Co-op staff	Consultant	Pubs & materials	Online course	Co-op council	Online course	Joint meet	Attend conf
					percent			
Pacific (n=60)	58.3	23.3	40.0	8.3	58.3	8.3	16.7	35.0
California (n = 28)	35.7	25.0	32.1	7.1	67.9	10.7	17.9	25.0
Washington (n = 20)	90.0	25.0	45.0	15.0	50.0	10.0	20.0	30.0
Intermountain (n=38)	65.8	36.8	26.3		52.6	10.5	18.4	31.6
Montana (n = 12)	50.0		41.7		41.7	8.3	16.7	8.3
Great Plains (n = 209)	57.9	27.3	36.8	8.6	70.8	6.7	12.0	32.1
Kansas (n = 47)	61.7	29.8	38.3	14.9	85.1	12.8	14.9	29.8
Nebraska (n = 33)	60.6	39.4	45.5	6.1	78.8	6.1	6.1	27.3
North Dakota (n = 41)	65.9	34.1	53.7	17.1	65.9	4.9	14.6	41.5
Oklahoma (n = 16)	68.8	25.0	25.0		87.5			25.0
South Dakota (n = 27)	48.1	3.7	22.2	3.7	37.0	7.4	18.5	29.6
Texas (n = 46)	45.7	23.9	28.3	2.2	67.4	4.3	10.9	34.8
Southeast (n = 46)	69.6	26.1	47.8	8.7	39.1	6.5	26.1	39.1
Tennessee (n = 23)	78.3	21.7	47.8	8.7	52.2	4.3	39.1	34.8
North Central (n = 262)	69.5	35.9	39.7	7.3	58.8	4.6	16.0	36.6
Iowa (n = 46)	71.7	43.5	43.5	13.0	82.6	10.9	23.9	45.7
Michigan (n = 12)	66.7	25.0	41.7		58.3		8.3	16.7
Minnesota (n = 72)	70.8	33.3	38.9	5.6	45.8	4.2	6.9	36.1
Missouri (n = 13)	76.9	53.8	30.8	7.7	38.5		15.4	23.1
Ohio (n = 18)	55.6	50.0	38.9		61.1		11.1	22.2
Wisconsin (n = 45)	68.9	44.4	48.9	6.7	73.3	2.2	24.4	46.7
Northeast (n = 21)	66.7	4.8	38.1	9.5	23.8	9.5	9.5	23.8
Pennsylvania (n=13)	76.9	7.7	46.2	15.4	30.8	15.4	15.4	30.8
Total (n = 636)	64.3	30.2	38.5	7.6	59.8	6.3	15.4	34.4

n = number that provided information.

[*] For actual wording of delivery methods on survey, see Table 10 footnote.

Table 13. Cooperatives' perceived level of need of education for audiences (rated on scale of 1-5 with 1 needing education the least, 5 needing it the most)

Audience	Average rating
Members	3.3
Directors	2.9
Employees	3.0
Management	2.6
General public	3.4

N = 710.

Table 14. Cooperatives' perceived level of need of education for audiences (rated on scale of 1-5 with 1 needing education the least, 5 needing it the most), by type

Cooperative type	Members	Directors	Employees	Mgmt	General Public
			average		
Farm supply (n = 328)	3.3	2.8	3.1	2.5	3.7
Grain and oilseed (n = 220)	3.4	2.9	3.1	2.5	3.4
Dairy (n = 42)	3.2	3.1	2.8	2.7	3.0
Fruits, veg. & nuts (n = 35)	3.0	3.4	2.9	3.0	2.6
Cotton & cotton gins (n = 33)	2.8	3.1	2.9	3.2	2.9
Livestock, poultry, & wool (n = 14)	3.3	3.1	2.4	2.6	3.0
Other marketing (n = 12)	2.7	4.1	2.7	3.4	1.5
Service (n = 26)	3.0	3.3	2.4	3.0	2.7
Total (n = 710)	3.3	2.9	3.0	2.6	3.4

n = number that answered corresponding question.

Table 15. Cooperatives' perceived level of need of education for audiences (rated on scale of 1-5 with 1 needing education the least, 5 needing it the most), by size

Cooperative asset size	Members	Directors	Employees	Mgmt	General Public
			average		
Less than $1 million (n = 60)	3.0	3.0	2.7	2.6	3.2
$1 million to $2.49 million (n = 67)	3.4	2.9	3.0	2.6	3.3
$2.5 million to $4.99 million (n = 122)	3.1	2.8	3.0	2.7	3.3
$5 million to $9.99 million (n = 131)	3.3	3.0	3.1	2.5	3.4
$10 million to $19.99 million (n = 139)	3.3	2.9	3.0	2.7	3.4
$20 million to $49.99 million (n = 110)	3.2	3.1	3.0	2.6	3.5
$50 million or more (n = 81)	3.4	2.7	3.1	2.5	3.6
Total (n = 710)	3.3	2.9	3.0	2.6	3.4

n = number that answered corresponding question.

Table 16. Cooperatives' perceived level of need of education for audiences (rated on scale of 1-5 with 1 needing education the least, 5 needing it the most), by location

Cooperative location	Members	Directors	Employees	Management	General Public
			average		
Pacific (n=65)	3.1	3.3	2.9	2.9	2.8
California (n = 31)	3.1	3.6	2.9	3.1	2.5
Cooperative location	Members	Directors	Employees	Management	General Public
			average		
Washington (n = 21)	3.2	2.8	3.2	2.6	2.8
Intermountain (n=46)	3.1	2.8	2.9	2.7	3.4
Montana (n = 16)	3.1	2.6	2.5	2.4	3.7
Great Plains (n = 232)	3.3	2.9	3.1	2.6	3.5
Kansas (n = 52)	3.3	3.0	3.0	2.4	3.3
Nebraska (n = 33)	3.3	2.9	3.0	2.5	3.8
North Dakota (n = 51)	3.4	2.7	3.2	2.6	4.1
Oklahoma (n = 18)	3.1	2.3	3.3	2.7	3.5
South Dakota (n = 28)	3.3	3.0	3.1	2.6	3.6
Texas (n = 51)	3.1	3.0	3.1	3.0	3.0
Southeast (n = 51)	3.2	2.9	3.0	2.2	3.7
Tennessee (n = 23)	3.7	3.0	3.2	2.0	3.8
North Central (n = 289)	3.3	2.9	3.0	2.6	3.4
Iowa (n = 49)	3.6	2.9	3.2	2.4	3.8
Michigan (n = 18)	3.3	3.4	3.1	3.2	2.4
Minnesota (n = 79)	3.1	2.8	2.8	2.6	3.3
Missouri (n = 19)	3.2	2.9	2.7	2.7	3.2
Ohio (n = 18)	3.3	3.0	2.6	2.4	3.1
Wisconsin (n = 46)	3.4	3.1	3.2	2.6	3.7
Northeast (n = 27)	3.7	2.9	2.4	2.3	2.9
Pennsylvania (n=15)	4.3	3.5	2.1	2.3	2.7
Total (n = 710)	3.3	2.9	3.0	2.6	3.4

States included in each region: Pacific—CA, OR, WA, AK, and HI; Intermountain—AZ, NM, CO, UT, ID, WY, and MT; Great Plains—ND, SD, NE, KS, OK, and TX; Southeast—AR, LA, MS, AL, GA, FL, SC, NC, TN, KY, VA, and WV; North Central—MN, IA, MO, IL, WI, IN, OH, and MI; Northeast—MD, DC, DE, PA, NY, NJ, CT, RI, MA, VT, NH, and ME.

States included are those that had more than 15 responses.

n = number that provided information.

Table 17. Cooperatives' perception of whether there are enough education resources (materials, programs, workshops, etc.) available to them

	Number	*Percent*
Yes	568	75.6
No	126	16.8
Didn't answer	57	7.6

N = 751.

Table 18. Cooperatives' perception of whether there are enough education resources available to them, by type

Cooperative type	Yes	No	Didn't answer
		percent	
Farm supply (n = 345)	76.8	16.5	6.7
Grain and oilseed (n = 225)	81.8	15.1	3.1
Dairy (n = 47)	55.3	27.7	17.0
Fruits, vegetables, & nuts (n = 39)	82.0	10.3	7.7
Cotton & cotton gins (n = 36)	75.0	11.1	13.9
Livestock, poultry, & wool (n = 16)	43.8	31.2	25.0
Other marketing (n = 17)	47.1	23.5	29.4
Service (n = 26)	73.1	19.2	7.7
Total (n = 751)	75.6	16.8	7.6

Table 19. Cooperatives' perception of whether there are enough education resources available to them, by size

Cooperative asset size	Yes	No	Didn't answer
		percent	
Less than $1 million (n = 75)	52.0	22.7	25.3
$1 million to $2.49 million (n = 74)	73.0	16.2	10.8
$2.5 million to $4.99 million (n = 125)	77.6	16.0	6.4
$5 million to $9.99 million (n = 139)	79.9	13.7	6.5
$10 million to $19.99 million (n = 142)	80.3	16.2	3.5
$20 million to $49.99 million (n = 113)	83.2	11.5	5.3
$50 million or more (n = 83)	71.1	26.5	2.4
Total (n = 751)	75.6	16.8	7.6

Table 20. Cooperatives' perception of whether there are enough education resources available to them, by location

Cooperative location	Yes	No	Didn't answer
		percent	
Pacific (n=68)	76.5	17.6	5.9
California (n = 33)	81.8	15.2	3.0
Washington (n = 21)	66.7	28.6	4.8
Intermountain (n=48)	64.6	25.0	10.4
Montana (n = 16)	62.5	31.3	6.3
Great Plains (n = 240)	80.4	13.8	5.8
Kansas (n = 53)	84.9	11.3	3.8
Nebraska (n = 34)	79.4	14.7	5.9
North Dakota (n = 53)	79.2	17.0	3.8
Oklahoma (n = 18)	66.7	27.8	5.6
South Dakota (n = 28)	82.8	13.8	3.4
Texas (n = 54)	79.6	9.3	11.1
Southeast (n = 58)	60.3	24.1	15.5
Tennessee (n = 24)	83.3	8.3	8.3
North Central (n = 303)	78.9	15.5	5.6
Iowa (n = 50)	82.0	14.0	4.0
Michigan (n = 18)	88.9	5.6	5.6
Minnesota (n = 84)	72.6	20.2	7.1
Missouri (n = 22)	59.1	31.8	9.1
Ohio (n = 19)	94.7		5.3
Wisconsin (n = 48)	81.3	14.6	4.2
Northeast (n = 34)	52.9	23.5	23.5
Pennsylvania (n=18)	44.4	33.3	22.2
Total (n = 751)	75.6	16.8	7.6

States included in each region: Pacific—CA, OR, WA, AK, and HI; Intermountain—AZ, NM, CO, UT, ID, WY, and MT; Great Plains—ND, SD, NE, KS, OK, and TX; Southeast—AR, LA, MS, AL, GA, FL, SC, NC, TN, KY, VA, and WV; North Central—MN, IA, MO, IL, WI, IN, OH, and MI; Northeast—MD, DC, DE, PA, NY, NJ, CT, RI, MA, VT, NH, and ME.

States included are those that had more than 15 responses to most questions.

n = number that provided information.

Table 21. Attainment of educational materials and programs

Resource	percent
USDA Cooperative Programs	42.8
State or regional cooperative council	58.1
Cooperative center	14.6
Cooperative development center	4.9
University	15.2
Online through a variety of sources	25.1
Other cooperatives	37.9
Other sources*	12.4

*Other sources include National Society of Accountants for Cooperatives (NSAC), attorneys, auditors, consultants, National Farmers Union, Farm Bureau, trade associations, etc.

N = 670.

Table 22. Attainment of educational materials and programs, by type

Cooperative type	USDA	Co-op* council	Co-op center	Dev.** center	University	Online***	Other co-ops	Other
				percent				
Farm supply (n = 300)	36.7	57.7	14.7	4.7	11.3	26.0	45.0	11.0
Grain and oilseed (n = 214)	43.0	66.8	16.4	6.5	18.2	22.4	35.0	8.9
Dairy (n = 40)	57.5	42.5	12.5	2.5	15.0	37.5	22.5	15.0
Fruits, veg. & nuts (n = 34)	52.9	38.2	5.9	5.9	11.8	41.2	32.3	26.5
Cotton & cotton gins (n = 33)	48.5	69.7	6.1		6.1	9.1	39.4	12.1
Livestock, poultry, & wool (n = 12)	83.3	41.7	16.7		50.0	33.3		33.3
Other marketing (n = 13)	46.1	38.5	30.8	7.7	7.7	15.4	38.5	38.5
Service (n = 24)	50.0	41.7	16.7	4.2	41.7	16.7	25.0	12.5
Total (n = 670)	42.8	58.1	14.6	4.9	15.2	25.1	37.9	12.4

n = number that provided information.

*State or regional cooperative council.

**Cooperative development center.

***On-line through a variety of sources.

Table 23. Attainment of educational materials and programs, by size

Cooperative asset size	USDA	Co-op council	Co-op center	Dev. center	University	Online	Other co-op	Other
					percent			
Less than $1 million (n = 50)	48.0	30.0	6.0	2.0	30.0	30.0	18.0	22.0
$1 million to $2.49 million (n = 61)	45.9	44.3	4.9	3.3	8.2	27.9	34.3	11.5
$2.5 million to $4.99 million (n = 111)	48.6	43.2	5.4	4.5	15.3	18.9	36.9	13.5
$5 million to $9.99 million (n = 125)	36.0	59.2	9.6	2.4	12.0	24.0	36.8	8.0
$10 million to $19.99 million (n = 135)	36.3	63.7	23.7	6.7	12.6	20.7	46.7	11.8
$20 million to $49.99 million (n = 108)	47.2	74.1	21.3	8.3	18.5	25.9	33.3	16.7
$50 million or more (n = 80)	45.0	73.7	23.7	5.0	16.2	36.2	47.5	7.5
Total (n = 670)	42.8	58.1	14.6	4.9	15.2	25.1	37.9	12.4

n = number that provided information.

Table 24. Attainment of educational materials and programs, by location

Cooperative location	USDA	Co-op council	Co-op center	Develop center	University	Online	Other co-op	Other
				percent				
Pacific (n=64)	50.0	42.2	9.4	7.8	9.4	28.1	39.1	23.4
California (n = 32)	50.0	25.0	15.6	9.4	9.4	34.4	21.9	25.0
Washington (n = 19)	36.8	52.6			10.5	15.8	73.7	26.3
Intermountain (n=40)	45.0	57.5	7.5	5.0	12.5	32.5	37.5	15.0
Montana (n = 12)	25.0	33.3		8.3	16.7	50.0	58.3	25.0
Great Plains (n = 222)	37.4	67.6	16.7	5.4	12.2	21.2	29.3	10.8
Kansas (n = 52)	36.5	76.9	44.2	13.5	19.2	17.3	17.3	7.7
Nebraska (n = 32)	31.3	90.6	6.3	3.1	12.5	34.4	28.1	3.1
North Dakota (n = 45)	40.0	60.0	17.8	4.4	6.7	24.4	33.3	13.3
Oklahoma (n = 18)	16.7	77.8	11.1		33.3	11.1	33.3	16.7
South Dakota (n = 28)	39.3	35.7		7.1	14.3	21.4	32.1	17.9
Texas (n = 48)	47.9	62.5	4.2		2.1	16.7	37.5	12.5
Southeast (n = 46)	52.2	58.7			26.1	23.9	32.6	13.0
Tennessee (n = 23)	43.5	73.9			17.4	17.4	30.4	4.3
North Central (n = 274)	42.7	55.1	18.2	5.1	15.3	26.6	47.4	9.5
Iowa (n = 46)	54.3	82.6	15.2	10.9	15.2	21.7	45.7	8.7
Michigan (n = 16)	56.3	18.8	18.8		12.5	31.3	43.8	18.8
Minnesota (n= 73)	31.5	50.7	13.7	6.8	15.1	30.1	53.4	9.6
Missouri (n = 17)	35.3	35.3		5.9	17.6	41.2	41.2	11.8
Ohio (n = 18)	44.4	72.2		5.6	33.3	22.2	55.6	
Wisconsin (n = 45)	42.2	62.2	57.8	4.4	11.1	28.9	37.8	8.9
Northeast (n = 24)	54.2	45.8	8.3		41.7	25.0	16.7	25.0
Pennsylvania (n=15)	53.3	26.7	6.7		33.3	26.7	13.3	40.0
Total (n = 670)	42.8	58.1	14.6	4.9	15.2	25.1	37.9	12.4

States included in each region: Pacific—CA, OR, WA, AK, and HI; Intermountain—AZ, NM, CO, UT, ID, WY, and MT; Great Plains—ND, SD, NE, KS, OK, and TX; Southeast—AR, LA, MS, AL, GA, FL, SC, NC, TN, KY, VA, and WV; North Central—MN, IA, MO, IL, WI, IN, OH, and MI; Northeast—MD, DC, DE, PA, NY, NJ, CT, RI, MA, VT, NH, and ME.

States included are those that had more than 15 responses to most questions.

n = number that provided information.

Table 25. Rating of types of education topics directors most need (scale of 1 to 5 with 1 needed the least, 5 needed the most)

Education topics Cooperative basics (what they are, principles,	Average rating
roles, practices, etc.)	2.8
Cooperative finance	3.4
Cooperative governance	3.0
Cooperative leadership	3.1
Cooperative board meeting functions	2.9
Cooperative tax and legal issues	3.2

N = 691.

Table 26. Rating of types of education topics directors most need (scale of 1 to 5 with 1 needed the least, 5 needed the most), by type

Cooperative type	Basics*	Finance	Govern.	Leadership	Meeting	Tax & legal
	average rating					
Farm supply (n = 317)	2.7	3.4	3.1	3.2	2.8	3.2
Grain and oilseed (n = 216)	2.7	3.3	3.0	3.1	3.1	3.3
Dairy (n = 40)	2.9	3.3	2.9	2.9	3.0	2.9
Fruits, veg. & nuts (n = 35)	3.3	3.1	3.0	2.7	2.8	3.0
Cotton & cotton gins (n = 32)	2.8	2.8	2.4	3.1	3.1	3.1
Livestock, poultry, & wool (n = 14)	3.4	3.2	2.7	2.9	2.6	3.3
Other marketing (n = 12)	2.9	3.6	2.8	3.4	2.9	3.5
Service (n = 25)	2.9	3.2	3.2	3.1	3.2	2.9
Total (n = 691)	2.8	3.4	3.0	3.1	2.9	3.2

n = number that provided information.

*See Table 25 for longer definition of column titles.

Table 27. Rating of types of education topics directors most need (scale of 1 to 5 with 1 needed the least, 5 needed the most), by size

Cooperative asset size	Basics*	Finance	Govern.	Leadership	Meeting	Tax & legal
	average rating					
Less than $1 million (n = 58)	3.1	3.1	2.3	2.6	2.8	2.8
$1 million to $2.49 million (n = 63)	3.1	3.2	2.7	3.1	2.9	3.3
$2.5 million to $4.99 million (n = 117)	2.8	3.0	3.0	3.1	2.6	3.0
$5 million to $9.99 million (n = 127)	2.8	3.5	2.9	3.2	3.1	3.1
$10 million to $19.99 million (n = 140)	2.8	3.5	3.2	3.3	3.0	3.2
$20 million to $49.99 million (n = 107)	2.6	3.5	3.2	3.1	2.9	3.3
$50 million or more (n = 79)	2.4	3.4	3.4	3.4	3.0	3.5
Total (n = 691)	2.8	3.4	3.0	3.1	2.9	3.2

n = number that provided information.

*See Table 25 for longer definition of column titles.

Table 28. Rating of types of education topics directors most need (scale of 1 to 5 with 1 needed the least, 5 needed the most), by location

Cooperative location	Basics*	Finance	Govern.	Leadership	Meeting	Tax & legal
				average rating		
Pacific (n=63)	3.2	3.0	2.9	3.1	3.3	2.9
California (n = 30)	3.2	3.0	3.2	3.1	3.2	2.8
Washington (n = 20)	3.3	3.0	2.9	3.3	3.2	2.8
Intermountain (n = 46)	2.6	3.7	3.1	3.0	3.2	3.1
Montana (n = 16)	2.4	3.5	3.0	2.8	2.1	3.9
Great Plains (n = 229)	2.7	3.3	3.0	3.2	2.9	3.3
Kansas (n = 50)	2.2	3.0	3.0	3.3	2.8	3.4
Nebraska (n = 33)	2.9	3.5	2.9	3.5	3.1	3.5
North Dakota (n = 50)	3.0	3.3	3.2	3.0	2.8	3.3
Oklahoma (n = 18)	2.2	3.1	2.6	3.2	2.9	2.8
South Dakota (n = 29)	2.5	3.6	3.3	3.1	2.9	3.4
Texas (n = 50)	2.9	3.2	2.7	3.0	3.0	3.2
Southeast (n = 48)	3.1	3.4	2.9	3.3	2.8	3.1
Tennessee (n = 23)	3.2	3.3	2.9	3.8	3.0	3.1
North Central (n = 280)	2.7	3.4	3.1	3.1	2.9	3.2
Iowa (n = 47)	2.9	3.7	3.5	3.3	3.1	3.6
Michigan (n = 18)	2.6	3.2	2.7	2.9	2.9	2.9
Minnesota (n = 77)	2.7	3.4	3.1	3.1	2.6	3.1
Missouri (n = 19)	3.2	3.0	2.8	2.7	2.2	2.9
Ohio (n = 18)	3.1	2.9	3.0	2.6	2.7	2.8
Wisconsin (n = 44)	2.6	3.7	3.3	3.7	3.3	3.3
Northeast (n = 25)	3.1	3.4	2.6	3.1	2.6	2.5
Pennsylvania (n = 14)	3.4	3.1	2.7	2.9	2.4	2.6
Total (n = 691)	2.8	3.4	3.0	3.1	2.9	3.2

States included in each region: Pacific—CA, OR, WA, AK, and HI; Intermountain—AZ, NM, CO, UT, ID, WY, and MT; Great Plains—ND, SD, NE, KS, OK, and TX; Southeast—AR, LA, MS, AL, GA, FL, SC, NC, TN, KY, VA, and WV; North Central—MN, IA, MO, IL, WI, IN, OH, and MI; Northeast—MD, DC, DE, PA, NY, NJ, CT, RI, MA, VT, NH, and ME.

States included are those that had more than 15 responses to most questions.

n = number that provided information.

*See Table 25 for longer definition of column titles.

Table 29. Selected findings for respondents that source cooperative councils, USDA, and other cooperatives for their educational resources

	Co-op councils (n=389)		USDA (n=287)		Other co-ops (n=254)	
Percent (#) that also use Cooperative Councils		---	61.3	(176)	54.3	(138)
Percent (#) that also use USDA resources	45.4	(176)		---	42.1	(107)
Percent (#) that also use other cooperatives	35.5	(138)	37.3	(107)		---
Percent that educate members		34.8		39.3		41.2
Percent that educate directors		90.9		83.5		88.2
Percent that educate employees		71.7		70.8		75.6
Percent that educate management		76.2		71.2		73.5
Percent that educate general public		14.7		16.9		16.8
Rating of member need of education*		3.3		3.2		3.3
Rating of director need of education		2.9		2.8		2.8
Rating of employees need of education		3.1		3.0		3.1
Rating of management need of education		2.6		2.7		2.5
Rating of general public need of education		3.4		3.3		3.5
Percent that think there are enough resources		87.4		86.8		81.9
Percent that think there are not enough resources		10.8		10.8		15.7
Percent that did not know if there are enough resources		1.8		2.4		2.4
Rating of director need of basics education		2.8		2.9		2.8
Rating of director need of finance education		3.4		3.3		3.4
Rating of director need of governance education		3.1		3.1		3.1
Rating of director need of leadership education		3.2		3.2		3.1
Rating of director need of board meeting functions education		2.9		3.0		2.9
Rating of director need of tax and legal education		3.2		3.3		3.2
Total sales ($ millions)		87.7		77.4		75.2
Net income ($ millions)		2.3		2.0		1.9
Total equity ($ millions)		13.1		11.5		11.8
Total assets ($ millions)		30.9		28.0		28.9
Full-time employees		81		75		75
Part-time employees		51		44		45

*All ratings based on a scale of 1 to 5, with 1 being least needed and 5 being most needed.

Table 30. Selected financial and other data for cooperatives, 2009

	Educators n=636	Non-educators n=115	All respondents N=751
		averages	
Total sales	$86,673,416	$31,467,630	$78,633,738
Total expenses	9,468,183	3,291,616	8,535,092
Net income	2,229,832	962,485	2,044,618
Current assets	20,261,534	8,032,475	18,409,143
Long-term assets	7,724,666	2,968,894	7,033,994
Total assets	32,425,833	12,409,255	29,393,831
Current liabilities	15,216,569	6,119,193	13,865,596
Total liabilities	20,552,397	7,400,088	18,601,919
Member equity	8,019,718	4,207,550	7,439,158
Total equity	12,119,860	5,271,117	11,081,057
Working capital	5,253,876	2,129,890	4,779,399
Return on assets	9.1%	6.4%	8.7%
Return on equity	17.3%	14.4%	16.9%
Current ratio	2.1	2.6	2.4
Debt to assets	46.0%	40.3%	45.2%
Sales to assets	3.0	2.9	3.0
Full-time employees	76	29	69
Part-time employees	41	10	36
Voting members	950	564	891

Appendix Notes

1. Survey—Education Initiative Questions

Farmer cooperatives were asked to answer six questions about how they practice cooperative education. These questions provided information on the types and extent of education that they are providing or promoting for their directors, members, managers, and employees.

- Does your cooperative provide training or educational materials concerning cooperative practices or issues to (check all that apply) members, directors, employees, management, or the general public?
- If yes, how is that education delivered (internally with cooperative staff, internally using a consultant, internally through use of educational publications and materials, internally through an online course, externally through a cooperative council or organization putting on a workshop, externally through an online course, joint

educational meeting with another cooperative, attendance of state or national conference where education is provided, other)?

- Rate the audiences of members, directors, employees, management, or the general public according to their need of cooperative education (1 needing education the least, 5 needing the most).
- Are there enough resources (materials, programs, workshops, etc.) for cooperative education (yes, no)?
- Where does your cooperative get its educational materials or programs (USDA, state or regional cooperative council, cooperative center, cooperative development center, university, online through a variety of sources, other cooperatives, other)?
- For cooperative directors, rate educational topics (cooperative basics, finance, governance, leadership, board meeting functions, tax and legal issues) as to their relative need (1 for topic that is least needed, 5 for most needed).

2. Survey Respondents versus Total Population of Farmer Cooperatives

This sample (respondents n = 751) appears to be representative of the total population of farmer cooperatives (n = 2,389) in the United States so it is very likely that the findings reflect the greater population to a significant degree. Evidence for this is shown in Appendix Table 1 which presents the number of respondent cooperatives by type, compared to the number of all farmer cooperatives by type. The percentages of cooperative types between the sample size and the total number of cooperatives are fairly similar.

Similar to the overall population, farm supply cooperatives make up most of the respondents' sample at 345 (total population is 970), and grain and oilseed cooperatives are second at 225 respondents (total population is 566).

3. Cooperative Type or Function

Type and function classifications are made according to the cooperative's major business activity. Marketing cooperatives derive most of their dollar volume from the sale of members' farm products. These were grouped into grain and oilseed; dairy; fruits, vegetables, and nuts; cotton and cotton gins; livestock, poultry, and wool; and other marketing (e.g., sugar, dry bean, rice, etc), depending upon which accounts for most of the cooperative's business volume. Farm supply cooperatives derive most of their business volume from the sale of farm production supplies. These cooperatives handle a wide variety of supplies, farmstead equipment, and building materials. Many also handle farm and home items such as heating oil, lawn and garden supplies and equipment, and food. Service cooperatives provide specialized business operations of farmers, ranchers, or cooperatives such as trucking, storing, drying, artificial insemination, livestock shipping, etc. Many cooperatives handle multiple commodities and provide both marketing and farm supply services, as well as facilities and equipment used to perform these services. These associations were classified according to the predominant commodity or function, as indicated by their business volume.

Appendix Table 1. Respondent population and total population of farmer cooperatives

	Respondent Population		All Farmer Cooperatives		Percent of Respondent to All
	Number	Percent	Number	Percent	
Farm supply	345	45.9	970	40.6	35.6
Grain and oilseed	225	30.0	566	23.7	39.8
Dairy	47	6.3	154	6.4	30.5
Fruits, vegetables, and nuts	39	5.2	187	7.8	20.9
Cotton and cotton gins	36	4.8	180	7.5	20.0
Livestock, poultry, and wool	16	2.1	126	5.3	12.7
Other marketing	17	2.3	124	5.2	13.7
Service	26	3.5	82	3.4	31.7
Total	751	100.0	2,389	100.0	31.4

References

Haskell, James. "Cooperative Education Can Help Renew and Revitalize Co-ops," *Rural Cooperatives,* U.S. Department of Agriculture, July/August, 2003. http://www.rurdev.usda.gov/rbs/pub/jul03/content. html

Wadsworth, James J. "Signs of Life. Have You

Checked Your Co-op's Education Pulse Lately?" Rural Cooperatives, U.S. Department of Agriculture, September/October, 2004. http://www.rurdev. usda.gov/rbs/pub/sep04/content. htm

Wadsworth, James J. "Cooperative Educators See Critical Need to Revamp, Expand Education Efforts," *Rural Cooperatives,* U.S. Department of

Agriculture, March/April 2009. http://www.rurdev.usda.gov/rbs/pub/mar09/ coop.htm

End Notes

[1] The International Co-operative Alliance, Statement of the Cooperative Identity: http://www.ica.coop/coop/principles

[2] "Co-op Educators See Critical Need to Ramp-up, Expand Education Efforts," Rural Cooperatives, March/April, 2009.

[3] A cursory examination of the data appears to indicate that the sample of respondents fairly well represents the total population of farmer cooperatives (see Appendix Note 2).

INDEX

I

J

L

M

O

P